Different Strategies of Housing Design

Edited by Ayşem Berrin Çakmaklı

Published in London, United Kingdom

IntechOpen

Supporting open minds since 2005

Different Strategies of Housing Design
http://dx.doi.org/10.5772/intechopen.77454
Edited by Ayşem Berrin Çakmaklı

Contributors
Silvia Lucciarini, Rossana Galdini, Kazutoshi Fujihira, Akubue Jideofor Anselm, Ivan Oropeza-Perez, Deborah Bowen, Lisa Quintiliani, Aysem Berrin Cakmakli

First published in London, United Kingdom, 2019 by IntechOpen
IntechOpen is the global imprint of INTECHOPEN LIMITED, registered in England and Wales, registration number: 11086078, 7th floor, 10 Lower Thames Street, London, EC3R 6AF, United Kingdom
Printed in Croatia

British Library Cataloguing-in-Publication Data
A catalogue record for this book is available from the British Library

Additional hard and PDF copies can be obtained from orders@intechopen.com

Different Strategies of Housing Design
Edited by Ayşem Berrin Çakmaklı
p. cm.
Print ISBN 978-1-78984-385-9
Online ISBN 978-1-78984-386-6
eBook (PDF) ISBN 978-1-83962-677-7

For EU product safety concerns:
IN TECH d.o.o., Prolaz Marije Krucifikse Kozulić 3, 51000 Rijeka, Croatia, info@intechopen.com or visit our website at intechopen.com.

Meet the editor

Dr. Ayşem Berrin Çakmaklı is an instructor in the Department of Architecture at METU. She received her B.Arch., M.S., and Ph. D. degrees from the Middle East Technical University, Faculty of Architecture, specializing in Building Science. Her master thesis was on sustainable architecture and sustainable building materials and their effects on human and environment health. The title of her Ph.D. thesis is "Life Cycle Assessment of Building Materials in Hotel Refurbishment Projects: A Case Study in Ankara". Her main fields of interest are ecological and sustainable architecture, energy efficiency, life cycle assessment of buildings, building performance evaluations, and lighting in architecture. She currently teaches energy and environmental design, appropriate building materials, building detailing modelling, and basic design studio.

Contents

Preface

In a perfect human settlement - providing food, shelter, communications, employment, an efficient environment for the occupants' better performance - all human needs are satisfied without significantly polluting or destroying the natural environment. The depletion and degradation of natural resources and their effects, diseases, and disasters must be taken into consideration. This is a global problem so there must be cooperation. The contents of this book are based on the different architectural strategies that aim to achieve a low-energy, healthy, and sustainable built environment using the extensive experience of the authors, researchers, and scientists in architecture, planning, and construction.

In addition to covering primer passive design principles in different climatic conditions through examples, the book devotes sections to different sustainability point of views: environmental, social, and economic. Determining the needed strategy according to function, economy, and occupant comfort and affordability is the crucial step. Creating a framework for designing and building sustainable and durable structures is important in order to reduce carbon emissions and operation costs. This book helps the reader achieve a sustainable development without destruction of the resources while also growing a universal awareness of protecting the living and non-living environment.

The book contains an introductory chapter and four chapters authored by several academicians and researchers in the housing field. Within every chapter, the reader will be given an overview of determined parameters on the subject and, in many cases, a description of the authors' original proposed solution. The content of the book is organized as follows:

Chapter 1: Introductory Chapter: Healthy and Affordable Housing Strategies

This chapter focuses on healthy housing due to the sustainable design strategies not having any harmful results on human and environmental health. This chapter shows that according to sustainability criteria, not only the environmental responsiveness but also social and economic responsiveness should be taken into consideration.

Chapter 2: Effects of Street Geometry on Airflow Regimes for Natural Ventilation in Three Different Street Configurations in Enugu City

In this chapter, efficient natural ventilation, according to the micro climate conditions of an urban environment, is analyzed. Two different regions (recirculation region and ventilated region) in an urban street have different characteristics because of the flow and the street geometry. This chapter looks at these regions of airflow cultivation in a high-density residential settlement in a tropical African city as a case study.

Chapter 3: Fundamentals of Natural Ventilation Design within Dwellings

This chapter presents the parameters that are intricate to occupant satisfaction; such as acoustical, lighting, indoor air quality, and thermal comfort. These occupant phase issues are generally ignored by the designers and constructers. It is stated in this chapter that natural ventilation is an excellent alternative to achieve an adequate indoor environment in a sustainable manner.

Chapter 4: Comprehensive Strategy for Sustainable Housing Design

Sustainable housing strategies, maximizing occupants' well-being, and minimizing the environmental load are the main concern of this chapter. In order to create a framework for this phenomenon, the author(s) generated a sustainability checklist and design guidelines. The chapter also presents a comprehensive visualization in order to adjust sustainable housing design to suit different and changing situations.

Chapter 5: Social Innovation and Environmental Sustainability in Social Housing Policies: Learning from Two Experimental Case Studies in Italy

The chapter discusses approaches and solutions developed by social housing in contemporary cities, using two Italian case studies. By analyzing housing strategies in a broader perspective in terms of its procedural dimension, the author(s) focus on two Italian social housing case studies. The important measures such as temporariness, flexibility, and the coordination are examined. The social, planning, architectural, and environmental quality of these two cases are represented.

Chapter 6: Socioeconomic Influences on Affordable Housing Residents: Problem Definition and Possible Solutions

The last chapter focuses on socioeconomic status (SES) as a powerful social determinant of health. After defining affordable housing, the author(s) discuss the future developments to improve the long-term health outcomes of these individuals and families due to survey data and geographical analysis. At the end they propose solutions to identify potential changes to affordable housing to promote

Summing up the issues presented in the book, it is directed at academic researchers, practitioners, and students in the housing and construction field. This book is the result of many collaborating parties. I gratefully acknowledge the assistance provided by **Gordon Tot and Ivan Butkovic**, Author Service Managers, and thank all the authors who contributed to this volume.

September 2019

Dr. Aysem Berrin Cakmakli
Middle East Technical University,
Ankara, Turkey

Section 1

Introduction

Chapter 1

Introductory Chapter: Healthy and Affordable Housing Strategies

Aysem Berrin Cakmakli

1. Introduction

In a satisfactory human settlement—providing food, shelter, communications, employment, and efficient environment for the occupants' better performance—all human needs are satisfied without seriously polluting or destroying the natural environment. All buildings must take place in the ecological cycle. Dynamic ecosystem can be formed from the relation with environment and buildings in a holistic view. However, the technological and economic development causes human beings to forget about or ignore ecological culture and buildings. As a result of this, one of the largest consumers of energy, housing sector, and its unconscious occupants' activities negatively affect the environment and conclude climate change. As highlighted in recent environmental studies, there are few areas in the world that can be considered "natural." Nearly half of the earth's surface is (in) directly affected by human interventions. Housing areas and transportation infrastructures have been prevailing in the development of a continuous and integrated green infrastructure in cities. Their critical contribution to urban climate strategies becomes undoubtedly evident when recent investments in both fields—construction areas and transportation—are considered. Since 1990, a variety of urban redevelopment and urban transformation projects, which introduced new residential and commercial areas, were added to the cities. The impacts of this changing urban form and density of the newly planned blocks on urban climate should have been investigated [1]. In order to avoid this, climate mitigation and adaptation strategies and energy-conservative and locally adaptive solutions should be applied to this sector. Architects and engineers have a major role in resolving the associated problems and maintaining comfort for occupants besides this. The contents of this book are based on the different architectural strategies that aim to achieve a low-energy, healthy, and sustainable built environment and its occupants through the extensive experience of the authors, researchers, and scientists in architecture, planning, and construction.

2. Sustainable housing strategies

Sustainable design includes resource and energy efficiency and healthier buildings and building materials that do not create building-related illnesses and supply-sensitive landscape. Among the architectural processes, flexible green site development and enhanced natural environment and resources are seen. Environmental responsiveness is applied to buildings by using such natural resources as the sun, wind, landforms, and natural vegetation to provide heating, cooling, lighting, ventilation, and protection from the elements. It is applied to infrastructure by capitalizing on natural features for water management, erosion

IntechOpen

control, and roadway design. It is possible for a new development not only to minimize damage to the local ecosystem but also to improve the surroundings. Sustainable building practices, the guidelines of reuse, reduce, and recycle to provide performance standards could be made concerning building design, elements, building materials, and energy and water systems. Besides this, sustainable design initiatives ensure environmental quality and performance of design by legislations and rating systems. The design should cut down the use of natural resources with a combination of energy-efficient technology and create less pollution than a standard production from the manufacture of the construction materials to the final work. Building orientation and its thermal mass capacity, natural ventilation, daylighting, and other passive strategies can lower a building's energy demand and increase the quality of the interior environment and comfort of the occupants [2]. The buildings must be designed to take advantage of the sun's energy and to avoid the summer's extreme heat. A tremendous amount of energy presently coming from nonrenewable sources can be saved. An example building of efficient energy use through proper design can be constructed with passive solar design, insulated buildings envelope, active low-energy systems, and active solar technologies. Using appropriate amount of fenestration to allow daylight, heat, and airflow into a building only when beneficial and arranging necessary natural or architectural shading devices to control heat gain are inevitable passive strategies to apply the designs. The process of supplying or removing air from an indoor space can be achieved by natural forces without the need of active or mechanical systems in different locations and climates; it is one of the main goal of passive housing strategies. On the other hand, these strategies in macroscale such as the geometry of buildings and periphery environment are also important parameters while designing. One of the main concerns of this book is explaining these parameters for urban environment in different climatic conditions through examples.

Bioclimatic architecture concerns the climate and environmental conditions to achieve thermal comfort inside of a building. To supply this, there should be perfect connection between design and natural elements such as sun, wind, rain, and vegetation, leading us to an optimization of resources. All new climate-adaptive dwellings will have to comply with many innovative techniques and specific architectural and technical criteria in order to reduce energy consumption through and have more comfortable dwellings. But it does not mean that there should be a complete dependence on mechanical systems, but they should be regarded as support. A good example of this is using natural ventilation or mixed mode ventilation [3]. The design of buildings and spaces should embrace and respond to the local climate by providing thermal and visual comfort, making use of solar energy and other environmental sources for heating, cooling, and lighting the buildings. Bioclimatic architecture makes the best use of the climate and the behavior of the occupants so as to reduce energy needs as much as possible. The key thing is to keep the inside of the dwelling comfortable without necessarily using air conditioning. One of the latest trends in architecture is to use new technology to enhance, amplify, and measure the performance of traditional bioclimatic techniques. Before being a slave of technical developments, the people inside older buildings were very good at adapting to climate to get the maximum performance [4].

Environmental education and being conscious through proper environmental development in designing activities and selecting building materials and products are also very important. Environmentally preferable building materials are durable and need low maintenance. Careful selection and specification can limit impacts on the environment and occupant health within the parameters of performance, cost, esthetics and availability. Modern constructions and surfaces due to the technological developments make human life easy in most areas as well as the environmental and human health problems they create [5]. While choosing surface materials, the framework of selecting building materials should be balanced environmental, social, and economic issues. Health and pollution should be of growing concern [6]. As buildings become more complex, liability and responsibility upon the shoulders of designers and to some extent the constructors of buildings also increase. Risks should be identified. This book is directed primarily for researchers, architects, construction planners, and building operators in order to make the proper selection from the design stage and take appropriate action to minimize environmental damage [7].

Besides environmental responsiveness, sustainability has another two main headings, economic sustainability and social sustainability. Although designers generally overlooked as an issue related to sustainable design, the impacts of the economic sustainability should be clearly measured and quantified. Just the opposite, social sustainability impacts are often intangible and difficult to measure especially at the scale of master planning and new community development. The contribution of the buildings to social sustainability is providing environments to meet social values like health and wellness, security and resilience, equity, and accessibility.

Determining the needed strategy according to function, economy and occupant comfort, and affordability is the crucial step. Affordable housing strategies are one of significant parameters of the health of individuals, families, and communities. The focus in contemporary domestic architecture seems to have shifted from human comfort to maximizing profits. Residential units are no longer designed to offer satisfactory accommodation; the number of rooms rather than the quality of space takes precedence, since the price of a flat is more dependent on the number of bedrooms rather than its area. Usually, kitchens are not large enough to accommodate all the functions associated with them, bedrooms are too small, and storage space is not even an issue. Poor accommodation conditions that expose mold, pests, and/or chemical toxins are harmful to human health. Besides physical effects, lack of affordable housing can be detrimental to the mental health of people living in low-to-moderate income, particularly for children and adolescents.

3. Conclusion

Buildings are not being designed according to user needs and preferences; the number of renovation and alteration projects has increased significantly during the past few years. The process of producing design guidelines should have these three steps: first, identifying environmental, social, and economic problems related to housing; second, determining the requirements to overcome these; and last, defining the desired goals for the satisfaction and well-being of the occupants and their environment. The problem is not only industrial and agricultural development processes but also the lack of adequate development. For developing and developed countries, the situation is different. Technology is needed for achievement of sustainable and environmentally sound development. But if it is not controlled, the nature is negatively affected for providing better economic situation. Developing countries should devise and adopt technologies which are better suited to their economic and environmental settings. For rural development, the generation and provision of energy should be based on solar potential and local resources. Creating a framework for designing and building sustainable and durable structures and environment is unavoidable in order to reduce carbon emissions and operation costs. This book makes the reader take enlightened decisions to achieve sustainable development without destruction of the resources with growing universal awareness of protecting the living and nonliving environment.

Author details

Aysem Berrin Cakmakli
Middle East Technical University, Ankara, Turkey

*Address all correspondence to: cakmakli@metu.edu.tr

References

[1] Milliga B. Corporate ecologies. Journal of Landscape Architecture. 2010;**5**(1):6-23

[2] Osso A, Walsh T, Gottfried D. Sustainable Building Technical Manual. New York: U.S Green Building Council Public Technology Inc; 1998

[3] Maciel AA. Bioclimatic integration into the architectural design [dissertation]. United Kingdom: University of Nottingham; 2007. Available from: https://pdfs.semanticscholar.org/3ef5/60bdd74654339a76d664e3f08fd0ecf0099f.pdf

[4] Blogactiv. Green Buildings: Bioclimatic Design, Passive Energy Systems and Renewable Technologies [Internet]. 2012. Available from: https://energypub.blogactiv.eu/2012/09/25/green-buildings-bioclimatic-design-passive-energy-systems-and-renewable-technologies/ [Accessed: September 2019]

[5] DPTI, Government of South Australia. Environmentally Sustainable Building Materials—Selection [Internet]. 2017. Available from: https://www.dpti.sa.gov.au/__data/assets/pdf_file/0009/293688/Environmentally_Sustainable_Building_Materials.pdf [Accessed: September 16, 2019]

[6] Ragheba A, El-Shimyb H, Raghebb G. Green architecture: A concept of sustainability. Procedia - Social and Behavioural Sciences. 2016;**216**:778-787

[7] Khatib JM, editor. Sustainability of Construction Materials. 2nd ed. United Kingdom: Woodhead Publishing/Elsevier Ltd.; 2016

Section 2

Passive Housing Design Strategies

Chapter 2

Effects of Street Geometry on Airflow Regimes for Natural Ventilation in Three Different Street Configurations in Enugu City

Jideofor Anselm Akubue

Abstract

Efficient natural ventilation is dependent on the micro climate conditions of an urban environment. This is affected by ambient wind flow, radiation and air temperatures. The airflow within the urban street can be cultivated into two regions. The first is a recirculation region, which forms in the near wake of each building. The Second is a ventilated region downstream of the recirculation region, formed when the street is sufficiently wide. The development of the flow into these two regions depends on geometry. This chapter looks at the impacts of street geometry on these regions of airflow cultivation in three different street configurations in high density residential settlements in Enugu city. It utilized schematic analysis of airflow regimes to identify the behaviors of flow in these street configurations relative to the height and width ratios of the street canyon. This schematic analysis can be utilized in preliminary design studies by city and building designers for justifying street dimensions and configurations in tropical regions where natural ventilation is paramount.

Keywords: urban street canyon, street geometry, street ventilation, natural ventilation, canyon effect

1. Introduction

Achieving natural ventilation is subject to the micro climate conditions of an urban environment. This is dependent on the ambient wind flow, radiation and air temperatures of the location. At the micro level, the urban space geometry is the most relevant parameter responsible for the microclimate variation [1]. Recent studies on urban air flow conditions identified the structure of wind field as characterized through the creation of vertical structure which occurs in two layers. When the air coming from the rural environments flows to the urban areas, it naturally adopts the new boundary conditions created by the city structure, thus resulting in the formation of the obstructed urban canopy sub layer, which is extended from the ground surface up to the buildings height, while the urban boundary free surface layer extends over the roof-tops. This urban canopy sub layer possesses specific flow properties which are defined by the interface between the air flow above the roof tops and the prevailing local conditions within the environment like the shape of buildings, street patterns, urban vehicular circulation, natural topography and the surrounding local landscaping structures like trees and shrubs. Wind speed in this canopy layer of the urban structure is generally decreased compared to the undisturbed normal wind speeds. Because of the reduction of the wind speed in urban street canyons, there is significant limitation to the extent natural ventilation can be applied in dense urban environments [2]. Designers of urban spaces are mostly unfamiliar with the thermal comfort problems arising from the relationship between urban forms and local climatological conditions [3]. Due to this, many urban settlements do not respond to contextual climatic conditions. The common practice of using universal meteorological data gotten from stations that are located in unobstructed areas of the city like the airports or isolated fields may result in inaccuracies regarding the air flow in typical urban buildings [4].

Recent trends in urbanization calls for sustainable ideas tailored towards improving thermal comfort. Understanding the flow pattern and wind speed in urban street canyons is therefore necessary for analyzing the natural ventilation potential of urban buildings, especially in the tropics which is characterized by high temperatures and humidity. The impacts of urbanization, the city heat island effects coupled with global warming contribute to the increase of the peak cooling loads and electricity consumption for space cooling purposes, owing to this and more, it is vital to optimize building and city designs in terms of energy demands that considers local urban microclimates.

2. Description of study

The potential for airflow for natural ventilation in urban streets is dependent on its configuration matrix. This study aims to examine the elements of street geometry that affect the flow system within the urban street. Firstly, it highlights the theoretical basis for the airflow studies in urban streets, it then identifies the primary influences, behaviors and parameters of flow within street canyons relative to natural ventilation. These parameters, which include the measurements of airflow recirculation and ventilation regions affected by height and width ratios, are believed to have major effects on the airflow regimes experienced in urban streets.

Using Enugu metropolitan area as a case, the study utilized schematic analysis process to examine three major street configurations that stand out in the residential districts. This process forms a basis for the analysis of airflow regimes and its effects on natural ventilation. Using the parameters mentioned above, each of these street configurations in the Enugu residential district is analyzed in order to justify the effective street size for natural ventilation in this region. The study identified the possibility of wind flow reduction in urban street canyons, because of this reduction, there exists a significant limitation in the extents natural ventilation may be utilized in dense urban settings like the tropical settlements in Enugu city.

Enugu city (6.26° North and 7.29° East) is located in the Southeastern part of Nigeria. **Figure 1** shows the climatic data of Enugu. Its climate is classified as a tropical rain forest with a derived savannah. The city is set on an altitude of 304.7 m above sea level, its climate is naturally humid and this humidity is regarded as oppressive between March and November. Its average maximum temperature stands at 34.9°C, with average lows of about 22.3°C and its annual mean temperature is about 26.7°C. Typical with the rest of West Africa, the rainy and dry seasons are the only weather periods in Enugu. The average hourly wind-speed in Enugu area experiences significant seasonal variations over the course of a year. For instance, the windier part

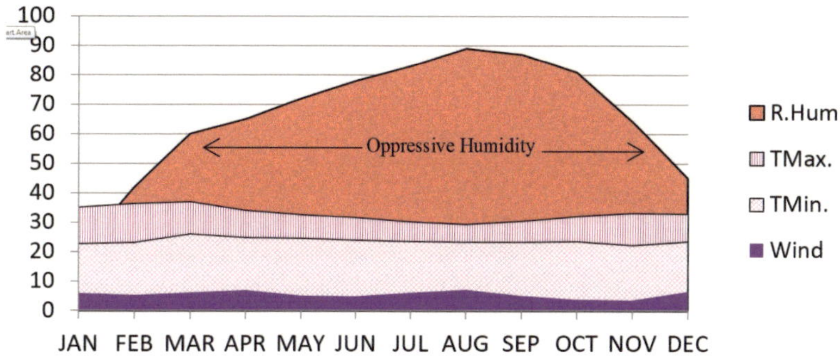

Figure 1.
Climatic data of Enugu City (source: NIMET year ended 2017).

of the year lasts for 7 months (between March and September), with average wind speed of 5.5 m/s. The windiest month of the year is usually August, with an average wind speed of 7.1 m/s. Typical wind path comes relatively from the west for an average of 10 months (termed the monsoon season), ranging from January to November, with peak percentage of about 75% occurring in July. Also, the wind direction changes to northerly for 2 months (termed the Harmattan season) from the months of November to January, with 41% occurring in the peak month of January. The wind data used in this study were obtained from the Nigerian Meteorological Agency (NIMET).

3. Elements of street geometry in urban settlements

The urban street has evolved from the original context as defined by the Latin word strata (meaning *"a paved road"*). In modern times, the function of the street in a built environment consists of more than the conventional roadway. Streets provide accessibility, connectivity, circulation, as well as infrastructural network for human settlements. As human activities grow, the function and configuration of a street system transforms especially in urban settlements. These transformations affect the micro climate environment within the street level as well as the universal urban environment. Urban climate is a critical factor which affects regional and global climates and consequently urban livability [5–7]. Geometry and orientation of an urban street affect outdoor and indoor environments, solar access inside and outside the buildings, the permeability to airflow for urban ventilation, as well as the potential for cooling of the whole urban system [8]. A typical urban street is comprised of stationary and mobile elements which form barriers that collectively affect the flow of wind within the streets. Owing to the presence of these obstacles like buildings, trees and street furniture, the air flow through the canopy layer is frequently obstructed compared to the air flow within the boundary layers. These results in the gentler airflow observed within an urban canopy layer compared to the airflow within adjoining rural settings [9]. **Figure 2** shows a typical urban street system and the barrier elements that affect airflow within the urban canopy layer.

3.1 Configurations of urban streets in Enugu city

Urban streets in tropical cities like the city of Enugu are structured to serve the basic conventional functions of accessibility and circulation. The configuration of Enugu streets is defined by urban planning classifications. The major classifications are the high density, medium density and low density. The low and medium density areas consist mainly of single villa residential housing units, while the high density areas consist of residential apartments. Three major residential districts stand out in these configurations (shown in **Figure 3**), these districts include:

i. Achara layout

ii. Independence layout

iii. GRA district

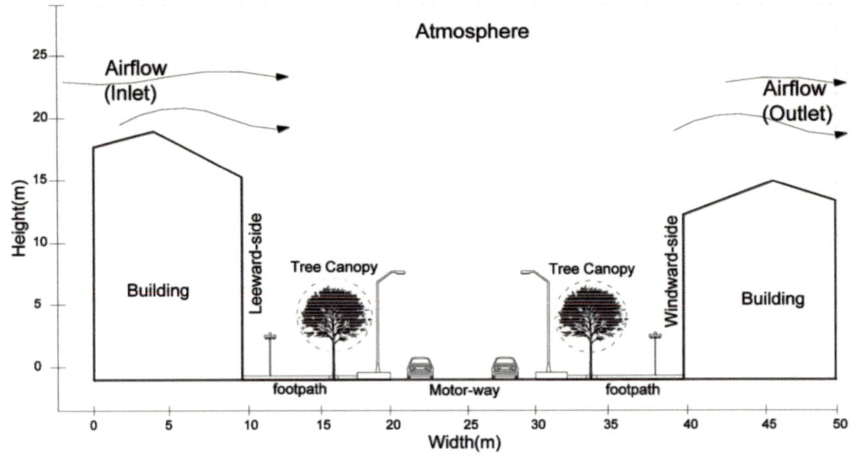

Figure 2.
Typical urban street showing the barrier elements that affect airflow within the urban canopy layer.

Figure 3.
Three major residential districts in Enugu city, representing the high, medium and low density districts.

Table 1 shows the configurations of the streets in Enugu relative to Height and Width ratios. Unlike the typical European streets characterized by activities like walkways, plazas, pavement cafes and other lifestyle facilities which define the size of the street outline (urban canyons), the sizes of the typical street canyon in Enugu city is more compact. This is condensed further by the presence of property boundary walls which is used mainly as solutions to the poor urban security.

3.2 Street canyon geometry

Street canyons are referred to as the space that is created within two similar rows of houses divided with a street way [10]. This configuration forms the major structure of modern city unit. Typical geometry of urban street canyons is defined by the characteristic ratio (aspect-ratio), which includes the ratio of the building heights *'H'* to that of the street width *'W'*. When a street canyon presents a common aspect-ratio that is approximately or equal to 1.0 without significant wall openings it is referred to as identical street canyon. When its aspect-ratio falls below 0.5, it is referred to as a shallow, while that with an aspect-ratio of 2.0 signifies a deep street canyon. The typical length of a street canyon *'L'* expresses the distance between two road intersections dividing the street into a short street canyon *'L/H = 3'*, medium street canyon *'L/H = 5'* and the elongated canyon *(L/H = 7)* [11].

Recent studies identified street geometry elements (height-to-width ratio *(H/W)*) and the street orientation as the most relevant urban parameters responsible for the microclimatic changes in a street canyon [12]. **Figure 4** illustrates the outline of a typical street canyon as parameters significant for major urban microclimate variations. These parameters directly affect the potential of airflow at street levels, solar access and its permeability and the overall urban microclimate [13, 14]. Consequently, designers and planners of urban streets are faced with the complexity of integrating seasonal climatic factors in the configuration and orientation of streets and buildings.

Configuration	Ratio	Planning classification	Street image	Aerial image
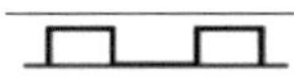	0.44	*GRA district* High rent residential district (bungalows and single home units)		
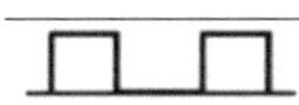	0.68	*Independence layout* High rent residential district (two floor homes made up of duplex and town houses)		
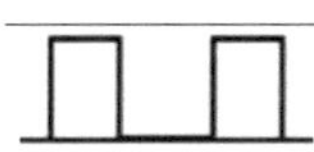	0.94	*Achara layout* Medium rent residential district (apartments and blocks of flats)		

Table 1.
Configuration of urban streets in Enugu city.

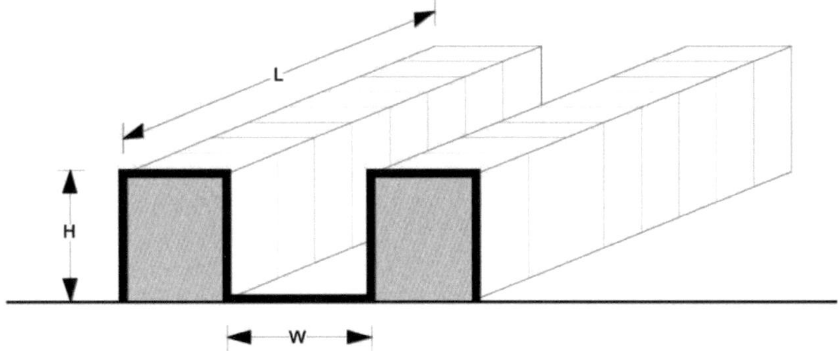

Where:
L = road distance between two main intersections subdividing the street canyon
H = height of the buildings
W = width of the street

Figure 4.
Outline of a typical street canyon.

For instance, in the tropical hot and humid regions, the major challenges are protection from the sun and permeability of natural ventilation, which naturally influence local designers' use of shallower and compact street canyons. However, the opposite is virtually the case in colder regions where solar access is required for winter seasons, hence necessitating more openness to the sky for buildings in that region.

4. Impacts of street configuration on airflow for natural ventilation

Due to the direct impacts of urban climate and built structures on the human thermal conditions, the control of outdoors microclimate is an important issue in urban designs. The integration of adequate, well-designed planning that places urban morphology in the climatic context of its environment, contributes to sustainable development of cities [3, 15–17].

Designers of such spaces like the architects, urban planners and engineers are often unfamiliar with the thermal comfort problems arising from the relationship between urban forms and local climatological conditions. According to reference [18], it is important to forecast the thermal consequences at the early stages of the design process of urban development areas prior to construction. In the case of humid tropical climates, the cooling effect of airflow particularly at nights could mitigate effects of urban heat island phenomenon. The significance of the formation of airflow within a street canyon is essential for human health and the energy efficiency of buildings and consequently, a pleasant urban microclimate [19]. These Urban airflow patterns are determined by the interaction between approaching wind within the built environment. Conversely, the pattern of an existing regional wind is changed when it flows through a built environment [20]. The air flowing over urban areas is sub-divided into two main layers.

i. The urban canopy layer

ii. The urban boundary layer.

The urban canopy layer is the layer flowing below the level of roof tops and within the spaces between buildings. This layer is usually influenced by solar energy falling on building facades and the ground. On the other hand, the urban boundary layer is the flow that is present above the average height of buildings. Within the urban boundary layer, heat transfer, pollutant emission, evaporation and transpiration and general contemporary urban development are the main factors that affect air conditions at this level [21]. Also, owing to the elements of street geometry (barriers as discussed above) airflow in the urban canopy layer is obstructed more than the airflow in the urban boundary layer. A secondary circulation feature driven by urban boundary layer provides airflow in an urban canyon which is strongly affected by the street orientation and geometry *(H/W and L/W)*. This circulation feature is referred to as Airflow Regimes.

4.1 Study of airflow in street canyons

The airflow within the urban street canyon can develop into two regions.

i. Recirculation region, which forms in the near wake of each building.

ii. Ventilated region (downstream of the recirculation region), formed when the street is sufficiently wide.

The development of the airflow into these regions depends on geometry, the different flow characteristics in these regions mean that the fluxes from these two regions scale differently. However, due to the complexity of building canopies, the flows through them are highly complex and site dependent [22]. **Figure 5** shows the scheme of airflow regimes.

From **Figure 6**, the recirculation region assumes a trapezoidal cross section [23]. Measurements show that the length of the recirculation region, *Lr*, is relative to the

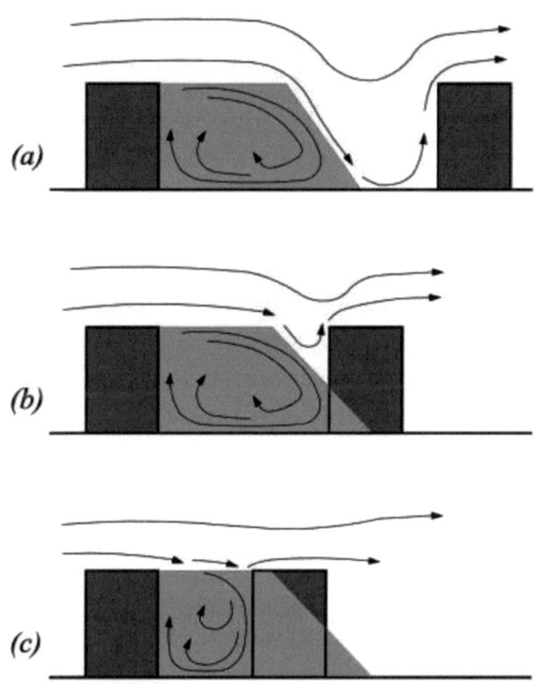

Figure 5.
Scheme of the three flow regimes in the urban boundary layer. (a) Lr < W (referred to as the isolated roughness flow regime); (b) Lr/2 < W < Lr (referred to as the wake interference flow regime); (c) W < Lr/2 (referred to as the skimming flow regime) [22].

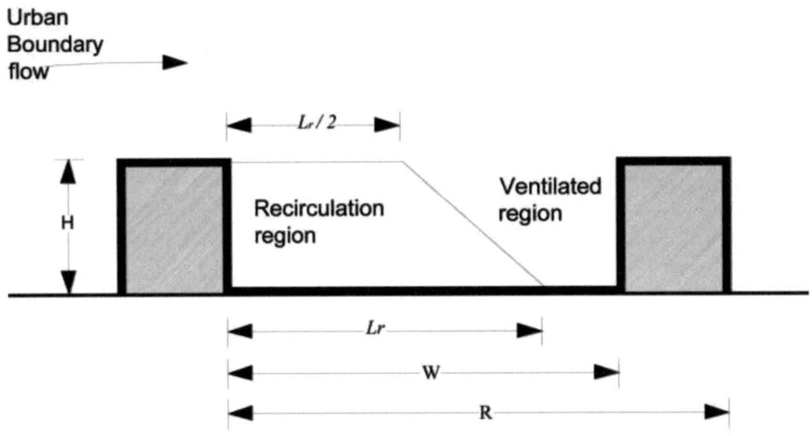

Figure 6.
Cross section of a street canyon showing the dimensions of the regions.

height of the building and so should be a scale relative to the height of the building. The value of the ratio *Lr/H* depends on the turbulence levels in the urban boundary layer and the shape of the buildings and roof.

Following the equations above, [22] suggests that *Lr/H* should be between 2 and 3. In the case of wide street canyons, it suggests that $Lr < W$, so that $H/W < 1/3$, in which case, the recirculation region does not impinge on the downstream building as can be identifies in **Figure 5a**. This it termed the "Isolated flow regime". In the case of intermediate (not so wide) street canyons, it suggests that $Lr/2 < W < Lr$, so that $1/3 < H/W < 2/3$, in which case, the recirculation region begins to impinge on the downstream building as observed in **Figure 5b**. This it termed the "Wake interference flow regime". Finally, in the case of narrow street canyons, it suggests that $W < Lr/2$, so that $H/W > 2/3$, in which case, the entire canyon volume is occupied by the recirculation regime as observed in **Figure 5c**. This it termed the "Skimming flow regime".

Inside the ventilated region, high-speed air flowing from above the roof top level is conveyed to street ground level. This causes the development of internal boundary flow layers along the surface of the street level. As this occurs, the vertical field of the wind regulates to a layer of even equilibrium with the normal surface of the street. However, the airflow in the recirculation region is dragged by recurrent high momentum injections of wind burst consistently drawn off the upstream roof. The steady burst of airflow slows as it advances around the recirculation region owing to the occurrence of gentler moving air induced by the associated street barriers [24].

4.2 Configuration of flow patterns in street canyons

As noted earlier, the conversion from one flow regime to another happens at critical combinations of *H/W* and *L/W* ratios. Understanding the configuration of patterns of airflow at different ratios is significant in determining the behaviors of flow regimes [25, 26]. **Figure 7** illustrates the characteristic interaction between the different flow regimes in urban canyons over building arrays of increasing *H/W* ratios. In the first instance, the isolated roughness flow regime occurs between buildings that are well spaced, in the event of zero interaction between the windward and leeward currents and similar to that of wind flow around isolated barriers. However, when the *H/W* ratio increases, the wakes are subjected to a disturbance, resulting to the formation of the wake interference flow. With additional increases of the *H/W* ratio the street canyon becomes isolated from the

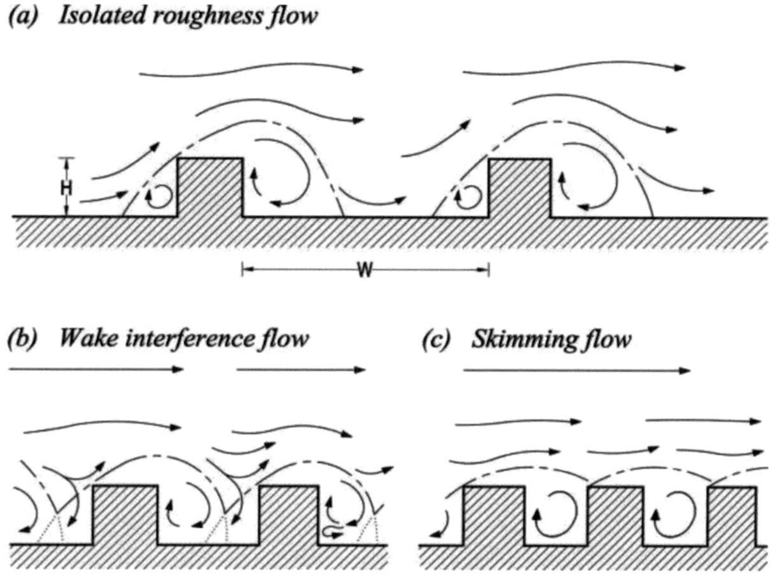

Figure 7.
The different flow regimes in urban canyons over building arrays of increasing H/W ratios [1].

flow of circulated air in the urban boundary layer, resulting in the creation of stable circulatory vortex within the canyon. This constant circulatory vortex produces the skimming flow regime that occurs mostly in the urban areas according to [27]. It could then be concluded that there is slower airflow in deep street canyons in comparison with intermediate or shallow ones.

4.3 Effects of street geometry on air temperature for natural ventilation

Studies by [28–31] show that when the wind speeds are small in urban streets, thermal effects come into play. Buildings tend to be warmer than the surrounding air and depending on the intensity of heating either the approach wind or the thermal convection dominate. The magnitude of the thermal change is dependent on many factors; including the anthropogenic heat flux in the canyon, sky view factor, and the albedo of the building facade.

Assessment of air temperature of street canyons in a hot climate in [32], show that the distribution of the air temperature recorded at the top and at the different floors of a building inside the canyon indicated that temperature above the canyon was higher during the day time than the associated temperature within the canyon. It was likewise observed that air temperatures were lower within the street canyon as the canyon is deep with limited permeation of solar radiation. This consequently results to lower surface temperatures especially towards the lower levels of the street which eventually brings about lower air temperatures. This lower temperature can be attributed to the mixing and the adjective phenomena inside the canyon that tend to regulate the air temperature.

From the study of solar access and dispersion in [1], a lower-limit of about 0.4 is assumed for ratio of *H/W* owing to the need to improve the degree of shelter for warmness and maintain a significant proportion of the temperature that is essential for the heat island warmth in colder climates. The upper limits of 0.60–0.65 ensure the maintaining of both atmospheric dispersion and solar access within the streets. Considering all factors to least a minimal extent, the most effective range of *H/W* ratio should then be between 0.4 and 0.6.

The ambient air temperature in street canyons influences the energy load of the building environment and determines the cooling potential of natural and hybrid ventilation techniques. Thermal phenomena as well as recurrent vortices which develop at the canyon corners usually dominate the air flow patterns inside the canyon. Since solar heating of the exterior facades system may create surface temperature differences that impacts on the air temperature conditions within the street canyon, such air temperature system may create local upwind along the warmer side of the facade and create a downward flow on the opposite side. In [33], clockwise vortices and counter-clockwise vortices were observed within the street canyon as the wind flow became almost perpendicular to the canyon axis, indicating that either the uninterrupted wind speed was high or that the thermal phenomena observed in the canyon corners contributed to the formation of the flow. This goes to show that the air temperature of the streets is mostly governed by more complex and regional factors than their ambient surface temperature despite the impacts of the local canyon geometry.

5. Analysis of airflow in the Enugu city street configurations

The analysis of the airflow patterns in the case of the residential streets in the tropical hot and humid Enugu city is presented using the flow regime analysis schematics for the different H/W configurations and assuming $Lr/H = 3$. The objective is to identify the extent of obstruction caused by existing street geometry on the airflow for Natural Ventilation in this region using schematic diagrams.

<table>
<tr><th colspan="4">Scheme</th><th>Result</th><th>Effective street width</th><th>Resultant airflow regime</th></tr>
<tr><td colspan="4">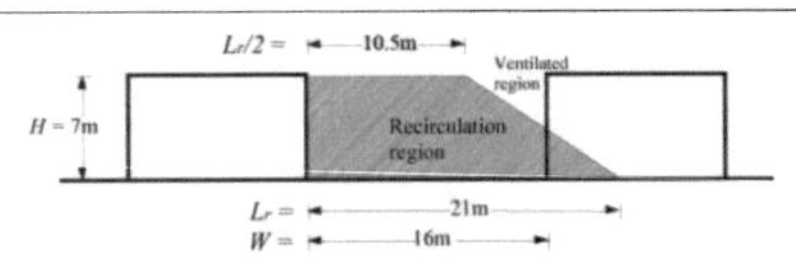
</td><td rowspan="3">Lr/2 < W < Lr
&
1/3 < H/W < 2/3</td><td rowspan="3">Intermediate-wide</td><td rowspan="3">Wake interference flow</td></tr>
<tr><td>H/W</td><td>Lr/H</td><td>Lr</td><td>Lr/2</td></tr>
<tr><td>0.44</td><td>3</td><td>21 m</td><td>10.5 m</td></tr>
<tr><td colspan="4">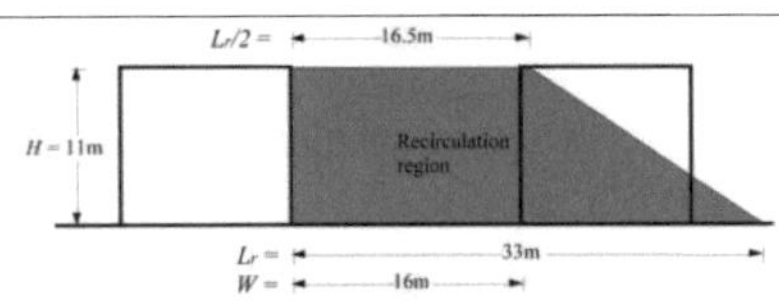
</td><td rowspan="3">W < Lr/2
&
H/W > 2/3</td><td rowspan="3">Narrow</td><td rowspan="3">Skimming flow</td></tr>
<tr><td>H/W</td><td>Lr/H</td><td>Lr</td><td>Lr/2</td></tr>
<tr><td>0.68</td><td>3</td><td>33 m</td><td>16.5 m</td></tr>
<tr><td colspan="4">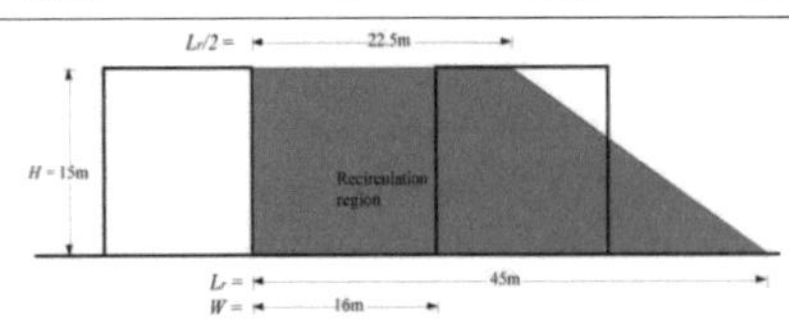
</td><td rowspan="3">W < Lr/2
&
H/W > 2/3</td><td rowspan="3">Narrow</td><td rowspan="3">Skimming flow</td></tr>
<tr><td>H/W</td><td>Lr/H</td><td>Lr</td><td>Lr/2</td></tr>
<tr><td>0.94</td><td>3</td><td>45 m</td><td>22.5 m</td></tr>
</table>

Table 2.
Schematic analysis of the airflow regime for the three main street configurations in Enugu city.

Figure 8.
Varieties of concrete walls used as boundary and party fence in Enugu city.

Configuration schematic	**Natural ventilation prospects**
Configuration 1 (0.44): GRA district	*Intermediate-wide street*
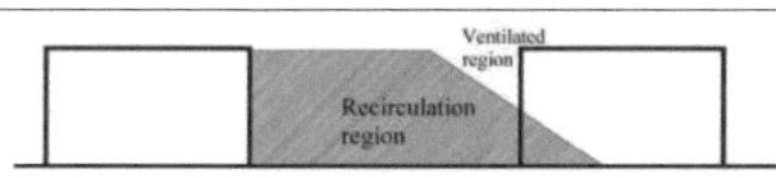	• Allows for continuous airflow from the urban boundary layer to the buildings (ventilated region), due to the lower heights of the buildings. • Unobstructed wind flow from urban boundary layer at the windward face of the buildings • Tendency of faster airflow for Natural ventilation
Configuration 2 (0.68): Independence layout	*Fine-narrow street*
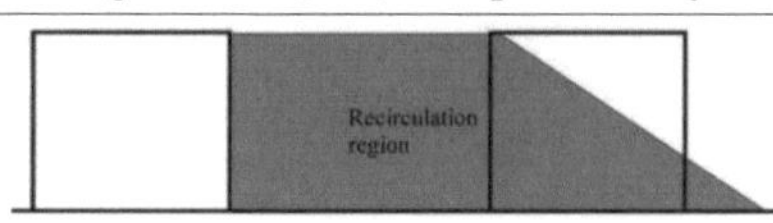	• Allows airflow mostly at the urban canopy layer • Wind flow is skimming type as it is mostly re-circulated airflow. • Slower winds lack the speed for adequate permeability of Natural ventilation • Natural ventilation in buildings can be improved by applying openings on the un-shaded faces in the schematic (ventilated region)
Configuration 3 (0.94): Achara layout	*Narrow street*
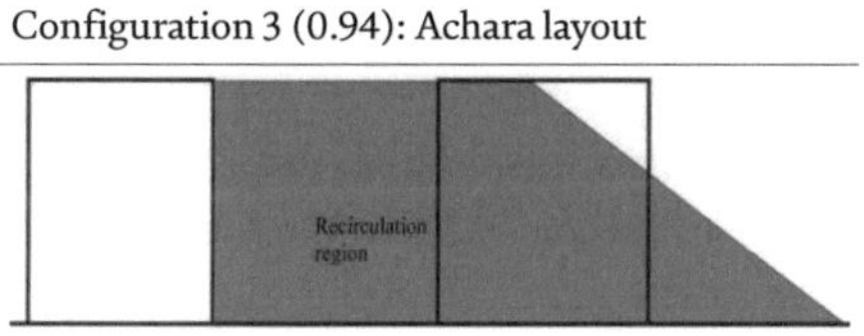	• Allows airflow mostly at the urban canopy layer • Wind flow is skimming type as it is mostly re-circulated airflow. • Slower winds lack the speed for adequate permeability of Natural ventilation • Natural ventilation in buildings can be improved by staggering buildings to break the continuous street canyon structure

Table 3.
Natural ventilation prospects for the three different street configurations.

The three different configurations observed and shown in **Table 1** have H/W ratios of 0.48, 0.68 and 0.94, respectively. For the 0.48 streets, the street width is approximately 16 m and the average building heights is approximately 7 m. In the case of the 0.68 streets, the street width is about 16 m and the average building heights is roughly 11 m. The 0.94 streets have street width of about 16 m and the average building heights is approximately 15 m. The schematics of the airflow regime for these streets are shown in **Table 2**.

From this study, it could be noted that the isolated roughness regime, which is the airflow that takes place between well-spaced buildings is not readily present in the urban streets of Enugu. From the schematic analysis above, the intermediate-wide street category only showed the potential to allow the continuous airflow from the urban boundary layer to the buildings, due to the lower heights of the buildings. This means that the speed of the airflow from the urban boundary layer is not disturbed at the windward face of the buildings, therefore permitting ventilated airflow. However, this potential is diminished by the presence of continuous often solid concrete fence walls with average heights of about 2.4 m (see **Figure 8**) which is often employed as security measures in buildings in this city. The other two categories in the Skimming flow regime are entirely dependent on airflow at the urban canopy layer which is the skimming airflow. This is due to the narrower street widths resultant from the building heights. Characteristics of skimming airflow include slower wind speeds and re-circulated airflow which lacks the speed for adequate permeability of natural ventilation in the buildings within these streets. The use of concrete fence walls is also rampant in this category of streets in Enugu which impedes further on the speed of the skimming flow. It is noteworthy however, that the orientation of a street, building elements (barriers) and prevalent wind directions ultimately affects the permeability of airflow for urban ventilation [8].

Table 3 presents the natural ventilation prospects in each of the three configurations shown above. With these scenarios, natural ventilation considerations could be carried along during the processes of building and urban street designs. However, the effectiveness of utilizing natural ventilation for different locations is improved through the orientation of streets and building planning.

6. Discussion and conclusion

A better understanding of the airflow and thermal characteristics within urban streets is paramount for developing urban area with energy saving considerations and human comfort. The rapid expansion of cities creates larger areas of increased drag, roughness and thermal forces that can significantly impact the flow patterns on regional scale. Hence, it is important to provide evaluation models with appropriate description of airflow, thermal conditions and natural ventilation models within tropical regions, as a function of meteorological frameworks that will accommodate predictions of airflow patterns on urban scales.

With climatic conditions prevalent in high temperatures and humidity, tropical cities like Enugu should aim towards the reduction of barriers and elements that impede on the airflow for natural ventilation. However, it is proper to reiterate that the major concern is with the comfort of occupants of the built environment. Similarly, identifying the criterion for judging the acceptable street climate conditions is a difficult task, as many objectives are involved. Assuming the major objective is for the airflow within street canyons, computations are available to calculate the effect of winds on the comfort and safety of humans in urban streets, for example see [34]. With knowledge of the general wind and temperature climatology and the effects of geometry, standard threshold of acceptable conditions can be predictable for any given location. However, the crucial objective is to establish airflow factors as a function of H/W in respect to natural ventilation within street canyons. Study identified the availability of wind tunnel and numerical models (as in [25, 35–37]) which can be applied with significant validation.

Through the analysis of the airflow regime for the three main street configurations in the tropical Enugu city, it could easily be concluded that the objective solution is to design streets with wider widths (i.e. smaller H/W ratios). A wider street provides better mixing of air and consequently better airflow in the urban street canyon. But this can never be the case in most modern city developments owing to land use objectives. Consequently, the other option is the provision of adequate openings between streets and courts and adopting different building heights in the streets which could improve air exchange within the urban canopy layer.

Author details

Jideofor Anselm Akubue
Architecture Department, Baze University, Abuja, Nigeria

*Address all correspondence to: akjideofor@yahoo.com

References

[1] Oke TR. Street design and urban canopy layer climate. Energy and Buildings. 1988;**11**:103-113

[2] Santamouris M. Energy and Climate in the Urban Environment. London: James and James Science Publishers; 2001. p. 412

[3] Gomez F, Perez A, Valcuende M, Matzarakis A. Research on ecological design to enhance comfort in open spaces of a city (Valencia, Spain); utility of the physiological equivalent temperature (PET). Ecological Engineering. 2013;**57**:27-39

[4] Geros V, Santamouris M, Tsangrassoulis A, Guarracino G. Experimental evaluation of night ventilation phenomena. Energy and Buildings. 1999;**29**:141-154

[5] Srivanit M, Kazunori H. The influence of urban morphology indicators on summer diurnal range of Urban climate in bangkok metropolitan area, Thailand. International Journal of Civil & Environmental Engineering. 2011;**11**(5):34-46

[6] Grimmond CSB, Roth M, Oke TR, Au YC, Best M, Betts R, et al. Climate and more sustainable cities: Climate information for improved planning and management of cities (producers/capabilities perspective). Procedia Environmental Sciences. 2010;**1**:247-274

[7] Huang LM, Li HT, Zhu DH. A fieldwork study on the diurnal changes of urban microclimate in four types of ground cover and urban heat island of Nanjing, China. Building and Environment. 2008;**43**:7-17

[8] Ali-Toudert F, Mayer H. Numerical study on the effects of aspect ratio and orientation of an urban street canyon on outdoor thermal comfort in hot and dry climate. Building and Environment. 2006;**41**:94-108

[9] Okeil A. A holistic approach to energy efficient building forms. Energy and Buildings. 2010;**42**:1437-1444

[10] Syrios K, Hunt GR. Passive air exchanges between building and urban canyon via openings in a single façade. International Journal of Heat and Fluid Flow. 2008;**29**:364-373

[11] Ahmad K, Khare M, Chaudhry KK. Wind tunnel simulation studies on dispersion at urban street canyons and intersections- a review. Journal of Wind Engineering and Industrial Aerodynamics. 2005;**93**:697-717

[12] Todhunter PE. Microclimatic variations attributable to urban canyon asymmetry and orientation. Physics and Geography. 1990;**11**:131-141

[13] Arnfield J, Mills G. An analysis of the circulation characteristics and energy budget of a dry, asymmetric, east-west urban canyon. Energy budget. International Journal of Climatology. 1994;**14**:239-261

[14] Nakamura Y, Oke TR. Wind, temperature and stability conditions in an east-west oriented urban canyon. Atmospheric Environment. 1988;**22**:2691-2700

[15] Kruger E, Minella F, Rasia F. Impact of urban geometry on outdoor thermal comfort and air quality from field measurements in Curitiba, Brazil. Building and Environment. 2011;**46**:621-634

[16] Taleghani M, Kleerekoper L, Tenpierik M, Van Den Dobbelsteen A. Outdoor thermal comfort within five different urban forms in the Netherlands. Building and Environment. 2015;**83**:65-78

[17] Wong PP, Lai P, Low C, Chen S, Hart M. The impact of environmental and human factors on urban heat and microclimate variability. Building and Environment. 2016;**95**:199-208

[18] Givoni B, Noguchi M, Saaroni H, Pochter O, Yaacov Y, Feller N. Outdoor comfort research issues. Energy and Buildings. 2003;**35**:77-86

[19] Memon RA, Leung DYC. Impacts of environmental factors on urban heating. Journal of Environmental Sciences. 2010;**22**(12):1903-1909

[20] Al-Sallal A, Al-Rais L. Outdoor airflow analysis and potential for passive cooling in the modern urban context of Dubai. Renewable Energy. 2012;**38**:40-49

[21] Thomas R, Fordham M, editors. Sustainable Urban Design: An environmental approach. London and New York: E & FN Spon; 2003. p. 46

[22] Oke TR. Boundary Layer Climates. London: Routledge; 1987. pp. 262-303

[23] Hertel O, Berkowicz R. Modeling Pollution from Traffics in a Street Canyon. Evaluation of Data and Model Development. Denmark: National Environmental Research Institute; 1989. Technical Report No.: DMU LUFT-A129

[24] Townsend AA. The Structure of Turbulent Shear Flow; Boundary Layers and Wall Jets. 2nd ed. London: Cambridge University Press; 1976. pp. 259-333

[25] Hussain M, Lee BE. An Investigation of Wind Forces on the 3D Roughness Elements in a Simulated Atmospheric Boundary Layer Flow; Part II- Flow Over Large Arrays of Identical Roughness Elements and the Effect of Frontal and Side Aspect Ratio Variations. London: University of Sheffield, BS; 1980. pp. 1-81

[26] Hosker RPJ. Flow around isolated structures and building clusters: A review. ASHRAE Transactions. 1985;**91**:1671-1692

[27] Yoshida A, Tominaga K, Watani S. Field measurements on energy balance of an urban canyon in the summer season. Energy and Buildings. 1990;**15-16**:417-423

[28] Tseng YH, Meneveau C, Parlange MB. Modeling flow around bluff bodies and urban dispersion using large eddy simulation. Environmental Science & Technology. 2006;**40**:2653-2662

[29] Kim JJ, Baik JJ. A numerical study of thermal effects on flow and pollution dispersion in urban street canyons. Journal of Applied Meteorology. 1999;**38**:1249-1261

[30] Fernando HJS. Fluid mechanics of urban atmospheres in complex terrain. Annual Review of Fluid Mechanics. 2010;**42**:365-389

[31] Sakakibara Y. A numerical study of the effect of urban geometry upon the surface energy budget. Atmospheric Environment. 1996;**30**:487-496

[32] Santamouris M, Papanikolaou N, Koronakis I, Livada I, Asimakopoulos DN. Thermal and air flow characteristics in a deep pedestrian canyon under hot weather conditions. Atmospheric Environment. 1999;**33**:4503-4521

[33] Georgakis C, Santamouris M. Experimental investigation of air flow and temperature distribution in deep urban canyons for natural ventilation purposes. Energy and Buildings. 2006;**38**:367-376

[34] Penwarden AD. Acceptable wind speeds in towns. Building Science. 1973;**8**:259-267

[35] Sievers U, Zdunkowski WG. A microscale urban climate model. Atmospheric Physics. 1986;**59**:13-40

[36] Nazarian N, Kleissl J. Realistic solar heating in urban areas: Air exchange and street canyon ventilation. Building and Environment. 2016;**95**:75-93

[37] Eliassona I, Offerle B, Grimmond CSB, Lindqvist S. Wind fields and turbulence statistics in an urban street canyon. Atmospheric Environment. 2006;**40**:1-16

Chapter 3

Fundamentals of Natural Ventilation Design within Dwellings

Ivan Oropeza-Perez

Abstract

Along with acoustical and lighting comfort, indoor air quality (IAQ) and thermal comfort upon households are essential to maintain a proper indoor environment, therefore ensuring a welfare toward the occupants. Nevertheless, sometimes, these features are neglected by building designers and constructers, causing problems such as the so-called sick building syndrome (SBS) and thermal discomfort, among others. Although there are short-term solutions such as purifiers, extractors, fans, and air conditioning, eventually these methods become not sustainable activities that consume energy and emit polluting gases such as chlorofluorocarbons. One alternative to this is natural ventilation, understood as the airflow throughout a building caused by changes of pressures naturally produced. In this chapter, the role of the early-stage building design as well as the correct occupant behavior is presented as essential to develop a naturally ventilated dwelling, which is an excellent alternative to achieve proper levels of indoor environment in a sustainable manner.

Keywords: natural ventilation, indoor air quality, thermal comfort, early-stage design, occupant behavior

1. Introduction

For the coming years, the so-called climate change will be a great challenge to be faced by the human kind, being the activities of this very same specie the most influencing factor for its rapid growth in the last decades.

One of these activities is the electricity generation, which in order to be produced, in many cases, a fossil fuel has to be burned. According to the International Energy Agency, "in 2016, generation from combustible fuels accounted for 67.3% of total world gross electricity production" [1] causing the emission of greenhouse gases and thus increasing the global warming.

In this context, according to the Intergovernmental Panel on Climate Change (IPCC), the building sector is an important target for the climate change because it accounts for almost 40% of the total energy use, embracing almost 25% of the global greenhouse gases emissions [2].

One of the most energy-consuming activities within the buildings is the space conditioning, in order to achieve proper levels of illumination, acoustics, indoor air quality (IAQ), and thermal comfort. For IAQ and thermal comfort, strategies of heating, ventilation, and air conditioning (HVAC) are applied by using fans, air-conditioning systems, and radiation systems, among others. These strategies, however, generally imply a high-energy consumption, including electricity. Moreover, IAQ and space cooling strategies use generally electricity for running fans, air-conditioning systems, evaporative coolers, air purifiers, air extractors, and air filters. According to Ekwall, only air-conditioning for cooling could account for more than half of the residential electricity usage of a single household located in a developing country with warm conditions [3].

Furthermore, if thermal comfort and IAQ are not achieved, especially in warm to hot countries, problems such as the sick building syndrome (group of symptoms of the occupants due to not-ventilated buildings such as dizziness, nausea, irritation, cough, itching skin, allergies, and headache [4]) and cold, flu, and other problems caused by high indoor temperatures could be arisen.

Therefore, strategies that do not imply high electricity consumption should be applied to achieve proper levels of IAQ and thermal comfort in warm to hot conditions. In this context, the usage of natural ventilation is proposed as a solution to these challenges while maintaining low levels of electricity use.

2. Fundamentals of natural ventilation

Natural ventilation within buildings is understood as the airflow throughout the construction propelled by pressure changes naturally produced. These pressure changes, or driving forces, are mainly two: wind pressure and buoyancy [5]. Both changes have the same physical origin but in different scales.

Wind pressure is defined as the wind incidence onto a building façade at a certain velocity and direction. When the wind impinges the façade, a positive pressure (or windward) occurs, while in the opposite façade of the building, negative pressure happens (leeward). An example of this can be seen in **Figure 1**.

Buoyancy, also called stack effect, is the air density difference between the outdoor and indoor air bodies caused by differences in the air temperature. It is the phenomenon when warm air "floats" among a colder mass of air. Although buoyancy is generally related to a vertical flowline, the flow could happen also in a horizontal direction. **Figure 2** shows how stack effect works.

As aforementioned, both driving forces have the same origin: temperature differences. In the case of buoyancy, the warm air is displaced toward a mass of colder

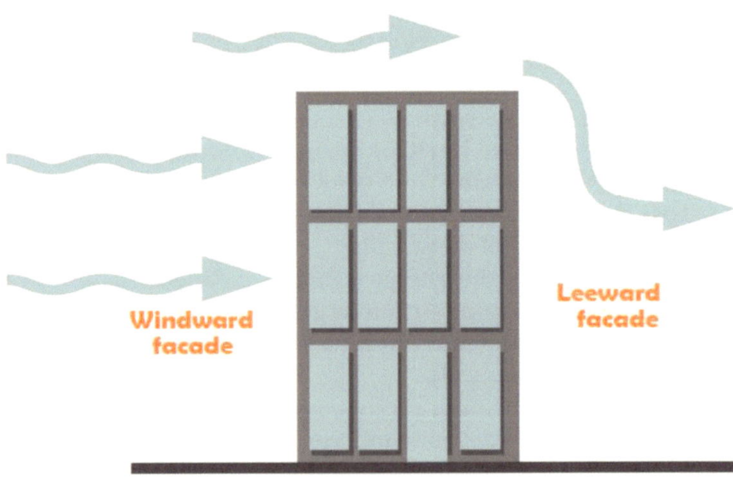

Figure 1.
Natural ventilation due to difference of pressure.

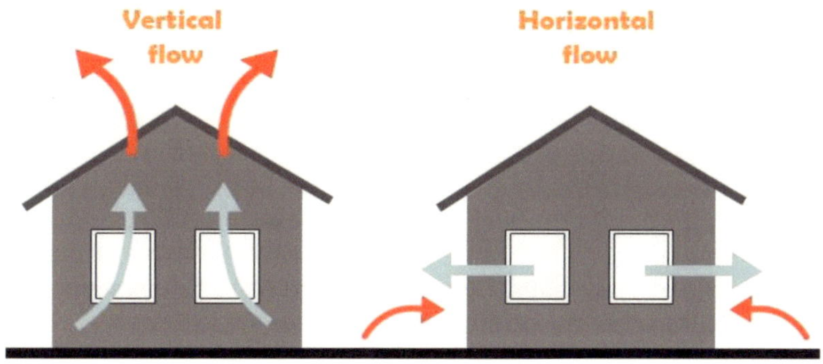

Figure 2.
Natural ventilation due to difference of temperature.

air, either in a vertical or horizontal direction. For wind pressure, the wind flow is headed due to the temperature difference occurred in a large scale. For example, when a mass of air is heated in some region, it floats up; therefore, a mass of cold air tries to cover the empty space that the floated air left. On the path to do this, the air hits the façades of the buildings, hence producing the airflow throughout them.

2.1 Natural ventilation for cooling and IAQ

As it was mentioned, natural ventilation can be used for cooling purposes as well as for achieving levels of IAQ. For the first subject, thermal comfort can be reached by three main manners:

- Cooling the indoor space as long as the outdoor air temperature is lower than the indoor air temperature.
- Refreshing the human body through evapotranspiration (removing the latent heat by the evaporation of the sweat drops of the skin).
- Cooling the building structure and internal heat gains by using convection heat transfer.

For having IAQ, natural ventilation is used by the displacement of the airflow in order to "sweep" the pollutants (positive pressure) or by "vacuuming" and expelling them (negative pressure).

3. Natural ventilation within buildings

Even though wind pressure might help to achieve thermal comfort and buoyancy might improve the IAQ, in this chapter, it is considered that stack effect only helps to reach thermal comfort, and wind pressure only improves the IAQ. This is because the temperature of the airflow is not always suitable to cool down a built space or the human body, in the case of wind pressure, whereas the airflow due to temperature differences is not always strong enough to expel all the pollutants from the indoor space.

3.1 Thermal comfort

The airflow rate by stack effect is calculated as follows [6–8]:

$$V_{stack} = (0.40 + 0.0045 \cdot \Delta T) \cdot A \cdot \sqrt{2 \cdot g \cdot H \cdot \left|\frac{\Delta T}{T_{indoor}}\right|} \quad (1)$$

where V_{stack} is the volumetric airflow due to buoyancy, ΔT is the difference between the outdoor and the indoor temperatures, A is the effective opening area (area of the opening where the airflow actually takes place), g is the gravitational acceleration (9.81 m/s^2), H is the height between the floor and the ceiling, and T_{indoor} is the indoor air temperature.

From Eq. 1 it can be seen that the volumetric airflow by buoyancy depends on the temperature difference, the ceiling height (for vertical airflow), and the effective opening area.

For cooling the space, the temperature difference should be negative; therefore, the indoor heat is expelled. For cooling the human skin, the outdoor temperature must be below 35°C; hence, the convective heat transfer from the human skin to its surroundings is negative (considering the temperature of the skin at approx. 37°C), and then, the latent heat of the skin is removed.

Thereby, if the outdoor conditions are suitable, thermal comfort can be achieved, either by cooling the space or the human skin. In this way, the two other terms of Eq. 1 can be varied in order to reach thermal comfort.

For this, in this document two types of approaches are proposed: building design at early stage and occupant behavior. For the first subject, the height of the ceiling must be correctly stated at the planning stage to optimize the airflow rate. For the second approach, a correct driving of the openings must be applied by the user depending of the indoor and outdoor conditions.

3.2 IAQ

Wind-driven ventilation is proposed to be carried out in order to remove the pollutants originated within the household. These pollutants (dust, odors, smoke, fungus, mist, fur, etc.) have their main origin in the bathroom and kitchen, although it can be generated at other rooms of the dwelling.

The model of volumetric airflow is shown below [6–8]:

$$V_{wind} = \left[0.55 + \left(\left|\frac{\alpha_{effective} - \alpha_{wind}}{180}\right| \cdot 0.25\right)\right] \cdot A \cdot v \quad (2)$$

where V_{wind} is the volumetric airflow due to wind, $\alpha_{effective}$ is the effective angle of the wind (set at 180°), α_{wind} is the actual angle of the wind, A is the effective opening area, and v is the wind speed.

As it can be noticed, Eq. 2 depends on the wind direction and speed as well as the opening area and the orientation of the façade (effective angle of wind). Thereby, the physical characteristics of the wind cannot be modified. Nevertheless, the orientation of the openings has to be analyzed before the dwelling is constructed, whereas the effective opening area could be driven by the occupants in order to adapt the airflow depending of their necessities.

Thereby, from Eqs. (1) and (2), three sets of inputs can be considered as the most influencing factors upon the natural ventilation performance. These sets are shown in **Table 1**.

In this way, the two approaches presented in this chapter are gathered for reaching thermal comfort and IAQ as follows: height of the ceiling and façade orientation as an early building design and a correct driving of the openings as a proper occupant behavior.

	Climate conditions	Building design	Occupant behavior
Stack effect	Difference of temperature	Ceiling height	Effective opening area
Wind pressure	Wind speed Wind direction	Building orientation	Effective opening area

Table 1.
Main factors for assessing the natural ventilation performance.

4. Natural ventilation proposals and performance

For assessing the performance of natural ventilation within a single dwelling, a model for stack effect ventilation and wind pressure ventilation is determined. A building of 120 m^2 of built area is considered with operable openings that can open from 0 to 3 m^2.

4.1 Thermal comfort

As aforementioned in this book chapter, ventilation by buoyancy is used to achieve thermal comfort within the dwelling, considering that the outdoor temperature is lower than the indoor temperature; therefore, indoor heat is removed. Another manner to reach comfort is the extraction of the indoor heat through a horizontal opening placed on the top of the building (see **Figure 2**).

Furthermore, in this book chapter the volumetric airflow, calculated with Eqs. (1) and (2), is seen as the performance of natural ventilation, either for achieving thermal comfort or for achieving IAQ. Hence, it is considered that the higher is the volumetric flow rate, the more efficient is the natural ventilation method.

Thereby, if an opening is placed in the top part of the building (ceiling/roof), an estimation of the volumetric airflow can be carried out. This architectural arrangement was chosen instead of a horizontal flow because the airflow would be more constant considering that the temperature difference is higher and more often than the occurred in a horizontal plane [5].

For an analysis of natural ventilation by stack effect, two parameters are considered: the ceiling height, which is taken as the building design, and the opening area, which is considered as a factor regarding the occupant behavior. **Figure 3** shows this airflow at different effective opening areas and different ceiling heights.

From **Figure 3** one can calculate that the volumetric airflow increases by approximately 0.12 m^3 per every 0.5 m^2 of the opening area. When the height of the ceiling is higher, the airflow rate increases as well. At the maximum opening area proposed here (3 m^2), there is a difference of almost 1 m^3/s between a height of 2.3 m and another of 2.8 m.

When the airflow is measured by changing the difference of temperature between the outdoor and the indoor air, **Figure 4** can be displayed. In this case, this factor is considered as climate condition, where there is no influence by the occupant behavior nor the building design.

In **Figure 4** it can be seen that the increase rate of airflow is lesser than the rate varying the ceiling height when the opening area enhances. Also, it is noticed that, at an effective opening area of 3 m^2, a temperature difference of 5°C makes a variation of only 0.3 m^3/s.

From **Figures 3** and **4**, it is noticed that the ceiling height, as long as it has an effective opening area, would help to expel the warm air and thus to reach indoor thermal comfort. The temperature difference, although it is an important parameter for the airflow rate, has a lower influence. Moreover, this physical characteristic cannot be controlled at any manner; therefore, its influence highly depends on the randomness of the outdoor and indoor conditions.

To achieve thermal comfort, however, two things must be considered: at an early-stage of construction, the ceiling height has to fulfill a certain value. From **Figure 4** one can establish that this value must be not lower than 2.5 m, if it is considered that from 2.5 to 2.8 m of height, the airflow rate can decrease up to 0.5 m^3/s.

Another approach to take into account is related to the opening driving, where the occupants have a total influence. For this, the openings placed on the ceiling/roof must be easily operable. Also, they have to be designed to protect the indoors from external hazards and conditions such as rain, animals, and burglars.

If these two approaches are correctly carried out, it is highly likely that the indoor temperature decreases. Furthermore, the correct driving of the openings would help to control the temperature if this value is lower or upper the comfort temperature range, set between 18 and 23°C, according to international standards [9].

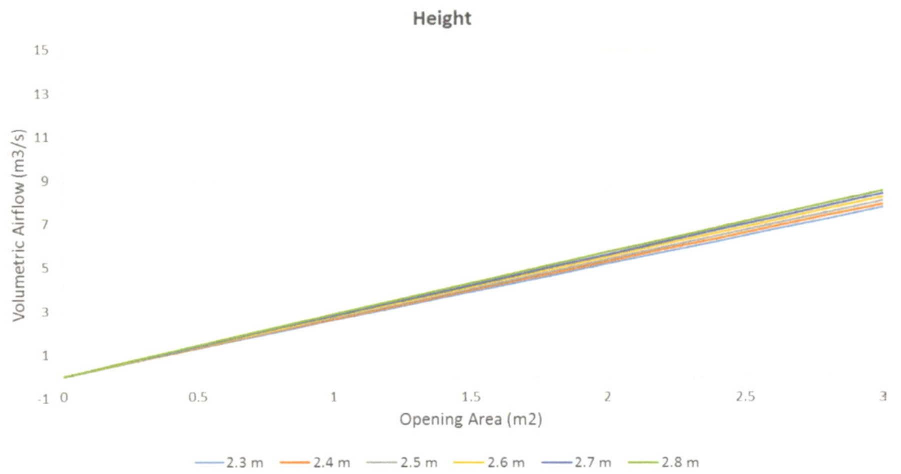

Figure 3.
Volumetric airflow by varying the ceiling height and the effective opening area.

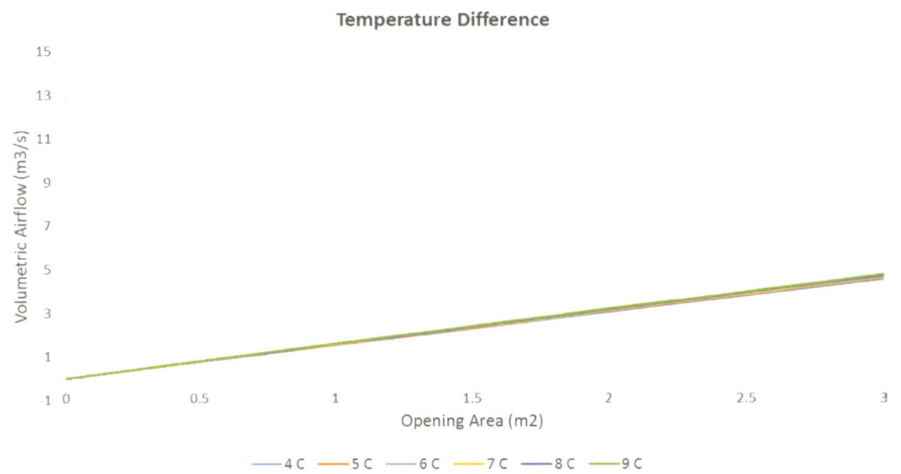

Figure 4.
Volumetric airflow by varying the difference of temperature and the effective opening area.

4.2 IAQ

For achieving correct levels of IAQ, cross ventilation is proposed. This type of ventilation is defined as a horizontal airflow through two openings placed on two façades opposed each other. Thereby, one opening has positive pressure, while the other has a negative pressure, hence increasing the one-direction airflow rate. Results varying the angle of incidence can be seen in **Figure 5**, where this factor totally depends on the building orientation, which is part of the building design.

In **Figure 5** the wind speed was set at 1 m/s, considering this value as sufficient to analyze the airflow rate by varying the angle of incidence. Thereby, it is noticed that the difference between an angle of 15° (almost parallel to the opening) and 90° (perpendicular to the opening) brings an airflow difference of 0.375 m^3/s. This means an airflow rate increases 0.05 m^3/s per every 10°. If the wind speed is higher, this airflow rate increases as well.

When the wind speed is varied, **Figure 6** can be displayed. In this case, the factor to vary, apart from the effective opening area, belongs to the climate conditions, which have no influence by the occupant behavior nor the building design.

In **Figure 6** the angle of incidence was set as perpendicular to the opening (90°). Results show a high airflow rate of increase by increasing the wind speed, having, at an effective opening area of 3 m^2, a difference of 12.15 m^3/s, being this the highest difference rate by varying the different features of ceiling height, temperature difference, angle of incidence, and wind speed.

Furthermore, as with the airflow by buoyancy, the wind speed is an outdoor characteristic that is not possible to control; thus, the airflow by wind pressure is highly dependent on its randomness.

Nonetheless, wind pressure airflow could consider two important features: at an early stage of design, the openings must be placed as most perpendicular as possible to the predominant wind direction of the place. This predominant direction could be taken from climate databases, very common nowadays for almost any place of the earth [10].

And as with stack effect airflow, the opening driving is a very important factor to considerer for controlling the airflow rate. As with the vertical flow by buoyancy, the horizontal airflow by wind pressure can be easily controlled by operable openings (vents, windows, doors) as long as the wind speed is sufficient to propel the airflow rate.

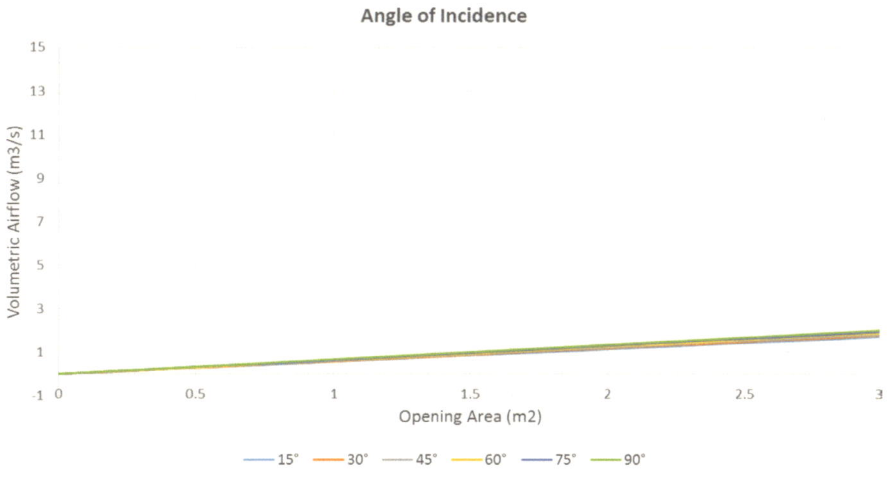

Figure 5.
Volumetric airflow by varying the angle of incidence and the effective opening area.

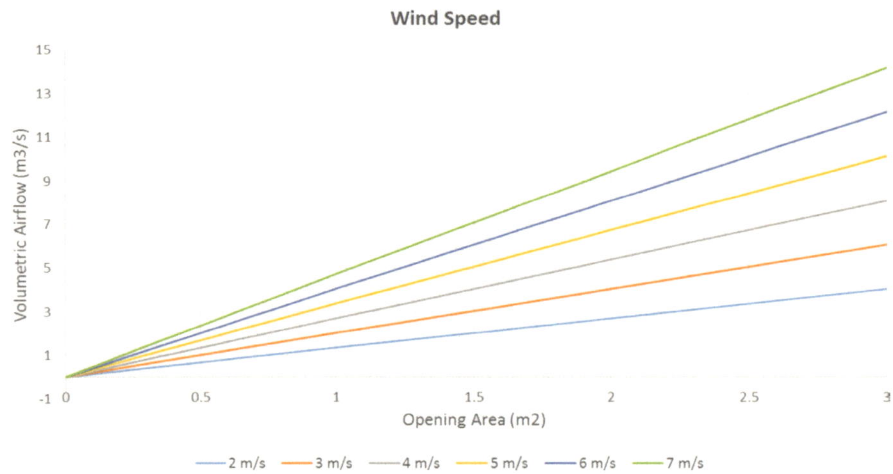

Figure 6.
Volumetric airflow by varying the wind speed and the effective opening area.

4.3 Natural ventilation proposals

With the prior analysis, two proposals are established as fundamental for natural ventilation within dwellings: stack effect for thermal comfort and wind pressure for IAQ. Also, three approaches for the two types of airflow are considered as follows:

- For buoyancy, a minimal ceiling height has to be taken into account. In this chapter 2.5 m is considered as enough.

- For wind pressure, one façade has to be perpendicular to the main wind direction of the place, which has to be calculated based on its local climate data.

- For both, operable openings must be placed. For the case of stack effect, these should be on the ceiling/roof or in the top part of the walls, whereas for wind pressure these should be placed at the façade perpendicular to the predominant wind direction and at its opposed façade.

The two proposals can work alone or together, considering that vertical airflow by stack effect is only efficient for warm conditions, where the temperature difference between the outdoor and the indoor air is high enough to propel the airflow. If the outdoor air temperature is low enough to cool down the indoor space, the only thing that the occupant has to do is opening the window, vent, or door, considering always that this occupant has the total control of the airflow by driving operable openings, either for buoyancy or for wind pressure.

5. Conclusions

In this book chapter, the fundamentals of natural ventilation for dwellings to achieve thermal comfort and indoor air quality are shown. Thereby, it is found that for thermal comfort, airflow by stack effect can be used, while for IAQ, wind pressure airflow could be applied.

In both cases, outdoor conditions, i.e., temperature difference between indoor and outdoor air and wind speed, are features that have high influence on the airflow rates. Nevertheless, these characteristics cannot be controlled by the building occupants; therefore, other approaches are considered.

These approaches are mainly three: the ceiling height, the orientation of the façades, and the driving of the openings. A comparison between **Figures 3** and **6** clearly shows the influence of these approaches upon the natural ventilation performance, where the ceiling height is the second most influencing parameter, followed by the building orientation. In both cases, the effective opening area plays a fundamental role to both increase the volumetric airflow rate and control it when it is necessary.

Nevertheless, in order to optimize the natural ventilation performance, various considerations must be taken. For the two first approaches, a correct design must be carried out before the construction of the dwellings, whereas for the correct driving, not only a proper design has to be done (operable openings), but the correct handling of the occupants is highly necessary.

If these approaches are correctly carried on, it is highly likely to achieve both IAQ and thermal comfort without spending any kind of resource, hence helping to the welfare of the occupants while countering the climate change.

Acknowledgements

The author wants to thank Dr. Astrid H. Petzold-Rodriguez for her useful comments regarding the clearness of the manuscript.

Conflict of interest

The author certifies that he has no affiliations with or involvement in any organization or entity with any financial interest (such as honoraria, educational grants, participation in speakers' bureaus, membership, employment, consultancies, stock ownership, or other equity interests and expert testimony or patent-licensing arrangements) or nonfinancial interest (such as personal or professional relationships, affiliations, knowledge, or beliefs) in the subject matter or materials discussed in this book chapter.

Nomenclature

$\alpha_{\text{effective}}$	effective angle of the wind entrance [°]
α_{wind}	wind direction [°]
ΔT	temperature difference between indoor and outdoor air [K]
A	opening area [m^2]
g	gravitational acceleration [m/s^2]
H	ceiling height [m]
T_{indoor}	indoor air temperature [K]
V_{stack}	volume flow rate due to stack effect [m^3/s]
V_{wind}	volume flow rate driven by wind [m^3/s]

Author details

Ivan Oropeza-Perez
Department of Architecture, Universidad de las Americas Puebla, Mexico

*Address all correspondence to: ivan.oropeza@udlap.mx

References

[1] International Energy Agency [Internet]. 2013. Available from: http://www.iea.org [Accessed: 2018-12-20]

[2] IPCC Working Group III. Special Report of Emission Scenarios [Internet]. 2000. Available from: http://www.ipcc.ch/publications_and_data/ar4/wg3/en/contents.html [Accessed: 2018-11-20]

[3] Ekwall T. Residential air conditioning in developing countries. Energy and Buildings. 1991;**17**(3):213-220. DOI: 10.1016/0378-7788(91)90108-F

[4] Ghaffarianhoseini A, AlWaer H, Omrany H, Ghaffarianhoseini A, Alalouch C, Clements-Croome D, et al. Sick building syndrome: Are we doing enough? Architectural Science Review. 2018;**61**(3):99-121. DOI: 10.1080/00038628.2018.1461060

[5] Etheridge D. Natural Ventilation of Buildings: Theory, Measurement and Design. 1st ed. John Wiley & Sons Ltd.; 2012

[6] ASHRAE Handbook of Fundamentals. New York, USA: American Society of Heating, Refrigerating and Air-Conditioning Engineers, Inc; 2001. Chapter 26

[7] ASHRAE Handbook of Fundamentals. New York, USA: American Society of Heating, Refrigerating and Air-Conditioning Engineers, Inc.; 2005. Chapter 27

[8] ASHRAE Handbook of Fundamentals. New York, USA: American Society of Heating, Refrigerating and Air-Conditioning Engineers, Inc.; 2009. Chapter 16

[9] Oropeza-Perez I, Petzold-Rodriguez AH, Bonilla-Lopez C. Adaptive thermal comfort in the main Mexican climate conditions with and without passive cooling. Energy and Buildings. 2017;**145**:251-258. DOI: 10.1016/j.enbuild.2017.04.031

[10] Han Q, Hao Z, Hu T, Chu F. Non-parametric models for joint probabilistic distributions of wind speed and direction data. Renewable Energy. 2018;**126**:1032-1042. DOI: 10.1016/j.renene.2018.04.026

Section 3

Assessment of Sustainable Housing Strategies

Chapter 4

Comprehensive Strategy for Sustainable Housing Design

Kazutoshi Fujihira

Abstract

Sustainable housing needs to be designed to maximize occupants' well-being and minimize the environmental load. The pursuit of combining these two different aspects toward sustainability is a goal-oriented task. The science of control can be applied to all goal-oriented tasks. Therefore, applying control science, we have been progressing in research on sustainable housing design. Our previous study has produced the control system for promoting sustainable housing design in which sustainable design guidelines and sustainability checklist are incorporated. Based on these accomplished results, this study has comprehensively visualized the process of producing and revising the sustainable design guidelines and sustainability checklist. Following this visualized process, also this study has concretely shown the production and revision processes of the sustainable design guidelines. The study results suggest that the comprehensive visualization can make these processes more manageable and help system designers to produce and revise the guidelines more efficiently. Furthermore, these results have led to indicating how to adjust the guidelines to different countries or regions as well as changing situations over time.

Keywords: sustainable housing, control system, sustainable design guidelines, sustainability checklist, production process, revision process, comprehensive visualization, communication

1. Introduction

Housing is the most important place that supports people's well-being. Sheltering residents from severe climate and weather, housing basically provides areas and facilities for sleeping, preparing meals, eating, and hygiene. People also conduct other activities, such as childcare, nursing care, communication, recreation, and learning, in their homes. Furthermore, the progress of information technology is currently increasing the number of people who work at home.

Meanwhile, housing is closely related to a variety of environmental problems. Typically, energy use at homes and the construction and demolition of houses have been significant causes of climate change, depletion of natural resources, and waste. Moreover, current and future impact caused by climate change is becoming a serious threat to human activities and people's well-being [1, 2]. Thus, this emerging threat requires housing design to strengthen the function of shelter.

Sustainable housing needs to be designed to maximize residents' well-being and minimize the environmental burden. The pursuit of combining these two different aspects toward sustainability is a goal-oriented task. The science of control can be applied to all goal-oriented tasks and has produced extraordinary results in many fields, especially engineering. Accordingly, applying control science, we have been progressing in research on sustainable housing design. In 2017, the already produced research results were compiled into the monograph titled *Sustainable Home Design by Applying Control Science*.

The core of the already produced results is the control system for promoting sustainable housing design where sustainable design guidelines and sustainability checklist are incorporated. Utilizing these research results, recently we have strived to clearly demonstrate the process of producing and revising the sustainable design guidelines and sustainability checklist. Moreover, following this demonstrated process, we have made the newest revision of the design guidelines. This chapter reviews these study results and discusses the meaning of the study.

2. Control system for promoting sustainable housing design

The "control system for promoting sustainable housing design" is demonstrated in **Figure 1**. The upper area of the figure is the theoretical world; the lower area is the practical world.

"Disturbances" are adverse effects on controlled objects which are caused by environmental, social, or economic problems. Concrete examples of the disturbances include harmful influences caused by environmental pollution and various impacts resulting from climate change. The route from "disturbances" to "sustainability" is "adaptation." This route has been added, based on the recent scientific understanding that adaptation measures to current and future impact caused by climate change are also necessary toward sustainability [1, 2, 5].

The purpose of control is the achievement of "sustainability." The model of sustainability (**Figure 2**) shows that sustainability needs both internal stability and fundamental stability, in order to realize the long-term well-being of all humankind or ultimate goal, within the finite global environment and natural resources or absolute limitations [6]. Internal stability means social and economic stability; the conditions for internal stability are "health," "safety," "mutual help," and "self-realization," which are important for the well-being of humans [6]. On the other hand, fundamental stability means environmental stability and a stable supply

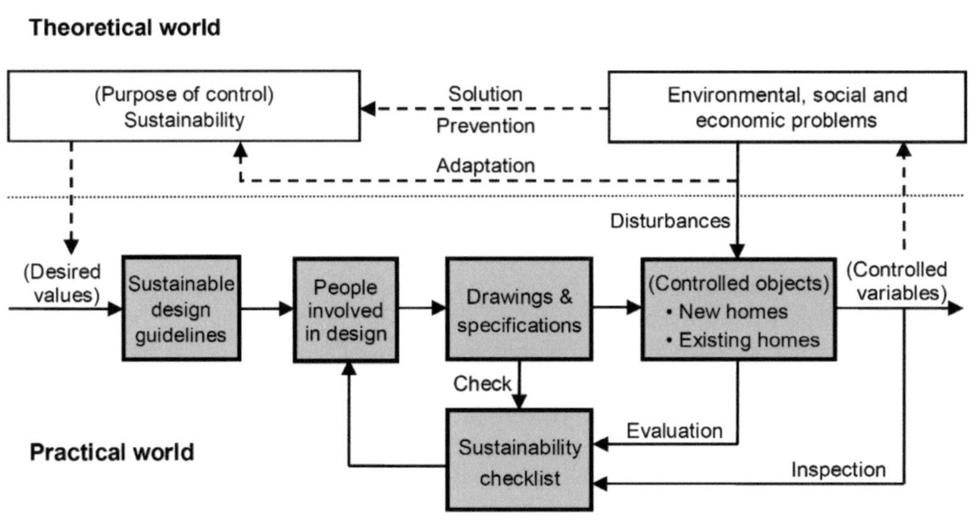

Figure 1.
Control system for promoting sustainable housing design [3, 4].

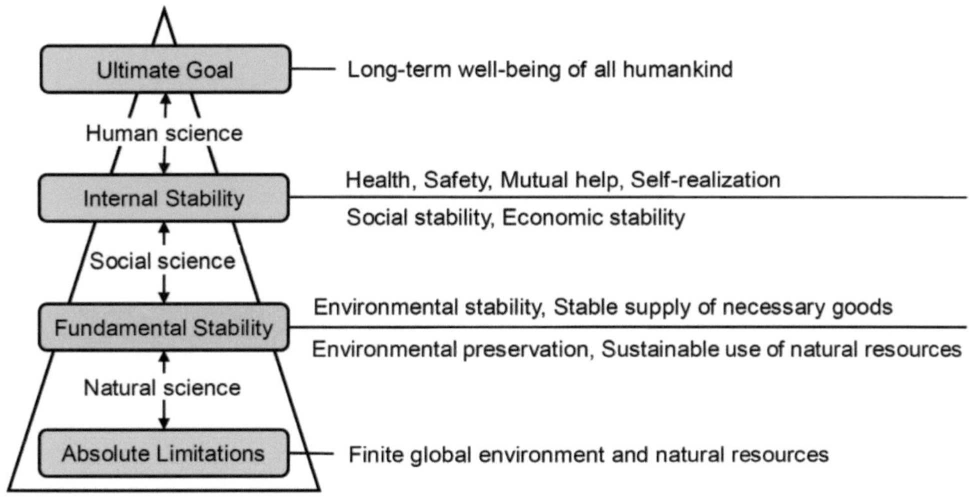

Figure 2.
Model of sustainability [6].

of necessary goods; the conditions for fundamental stability are "environmental preservation" and "sustainable use of natural resources" [6].

"Desired values" are derived from the purpose of control, namely, sustainability. Meanwhile, "controlled variables" are the variables that relate to controlled objects and need to be controlled for primarily solving or preventing the problems or adapting to disturbances [3, 4]. The control objective of this control system is to adjust the controlled variables to their desired values.

In the practical world, controlled objects are both "new homes" and "existing homes." "People involved in design," such as homeowners, architects, designers, and homebuilders, adjust the controlled variables to their desired values, by utilizing the "sustainable design guidelines" and "sustainability checklist." Both the design guidelines and checklist have almost the same structure, namely, elements, variables, and their desired values. However, the checklist is created to easily compare measured or estimated variables with the desired values and search for controlled variables [3, 4].

When objects are new homes, first, information on the desired values reaches "people involved in design" through the "sustainable design guidelines." People involved make "drawings and specifications," so that the variables of home's elements can attain their desired values as much as possible. At important steps in the design process, people involved check the drawings and specifications, by referring to the "sustainability checklist" [3, 4].

When existing homes are objects, the design process begins with "inspection" on the home as an object. The "people involved in design" measure or estimate each element's variables of that home by referring to the "sustainability checklist." After the inspection, the people involved usually make "drawings and specifications" for improvement, so that controlled variables satisfy their desired values as much as possible [3, 4].

3. Process of producing and revising the design guidelines and checklist

The process of producing and revising the sustainable design guidelines and sustainability checklist is shown in **Figure 3**. The upper and lower areas divided by the dotted line represent the "theoretical world" and the "practical world," respectively.

The central part demonstrates the course of preparing and using the "sustainable design guidelines" and "sustainability checklist." First, system designers produce

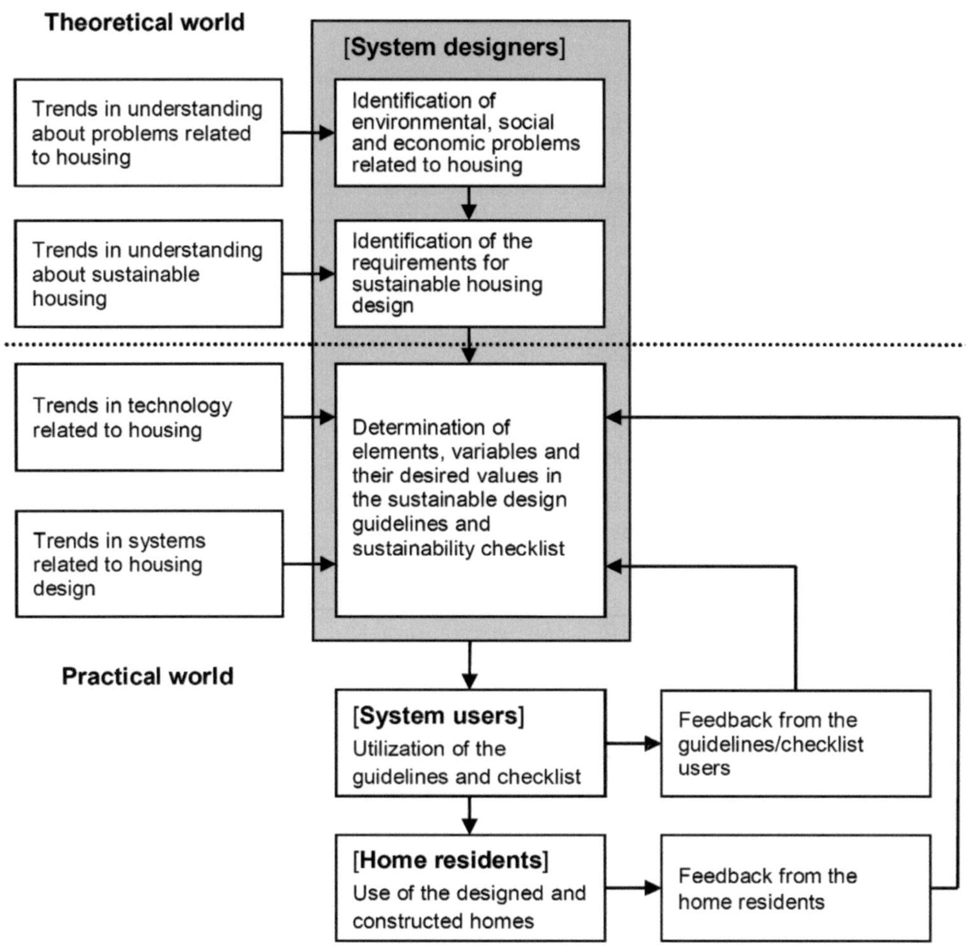

Figure 3.
Process of producing and revising the sustainable design guidelines and sustainability checklist.

or revise the guidelines and checklist through the three-stage process. Next, system users utilize the guidelines and checklist. After that, the residents use the actual homes that have been designed with the guidelines and checklist.

The four blocks on the left show the items to check when producing or revising the guidelines and checklist. The contents in these four items can change over time. Meanwhile, the two blocks on the lower right show the items to check when revising the guidelines and checklist, based on the feedback from the system users and home residents.

3.1 Process of producing the design guidelines and checklist

The process of producing the design guidelines and checklist consists of three stages: (1) identification of environmental, social, and economic problems related to housing, (2) identification of the requirements for sustainable housing design, and (3) determination of elements, variables, and their desired values in the design guidelines and checklist.

3.1.1 Identification of problems related to housing

Producing the guidelines starts with the identification of environmental, social, and economic problems related to housing. Observing trends in understanding about problems related to housing, system designers search for the problems that should be identified. The basis for the identification is that the problems affect the total six stability conditions, namely, health, safety, mutual help, self-realization, environmental preservation, and sustainable use of natural resources. When identifying problems, system designers take up both global/general problems and local/particular problems in their region or country. Examples of global/general problems are global warming and climate change, depletion of natural resources, and waste. Meanwhile, damage caused by earthquakes is included in local/particular problems.

3.1.2 Identification of the requirements for sustainable housing design

In the second stage, based on the identified problems related to housing, system designers identify the requirements for sustainable housing design. For example, if "depletion of natural resources" and "waste" are identified as problems in the first stage, the "extension of housing lifespan" and "use of resource-saving or waste-prevention materials" are usually identified as the requirements. These two requirements are related to "sustainable use of natural resources," one of the six stability conditions. Similarly, "damage caused by earthquakes" requires "higher resistance to earthquakes," which is related to "safety," another stability condition. In addition, observation of "trends in understanding about sustainable housing" also helps the system designers to identify such requirements.

3.1.3 Determination of elements, variables, and their desired values

In the third stage, the requirements for sustainable housing design are converted into the "element-variable-desired value" structure of the guidelines and checklist. The purpose of this conversion is the convenience of users in the practical world. The structure of "element-variable-desired value" shows design targets of each part of homes; therefore, it enables users to easily understand what should be designed and the courses of design.

3.1.3.1 Elements

The previous publications have presented a method of selecting important elements of homes, based on two main factors: "material" and "space." "Material" regards housing as the aggregate of material elements, such as framework, exterior, thermal insulation, windows and doors, interior, piping, and equipment for harnessing natural energy. "Space" considers housing as the aggregate of spatial elements, such as rooms and areas [3, 4]. In addition, material elements and spatial elements are equivalent to actual parts of homes. Accordingly, when designing, checking, evaluating, or inspecting homes, system users can easily compare the guidelines and checklist with the actual home or drawings [3, 4].

When selecting elements, system designers also consider the requirements for sustainable housing design. In other words, the elements that are required for sustainable housing design need to be included in the set of elements. For instance, "equipment for rainwater use" should be selected as one of material elements, even though it is not common in current ordinary homes [6].

3.1.3.2 Variables

After selecting elements, system designers examine the relationships between each element and the relevant stability condition(s), as well as the related requirement(s) for sustainable housing design. Subsequently they determine the element's variables that can indicate the degree of stability [3, 4].

Choosing only one element, namely, "framework," the rest of this part shows how to determine the variables. System designers examine the relationships between "framework" and "sustainable use of natural resources," the relevant stability condition, as well as the "extension of housing lifespan" and "use of resource-saving or waste-prevention materials," the related requirements. As a result, they can identify "durability" and "materials," indicators of "sustainable use of natural resources," as variables of "framework." Moreover, if the country or region is in an earthquake-prone area, the system designers consider the relationships between "framework" and "safety," the relevant stability condition, as well as "higher resistance to earthquakes," the related requirement. They can subsequently identify "resistance to earthquakes," an indicator of "safety," as an additional variable.

3.1.3.3 Desired values

After determining variables, system designers set the desired values of these variables that can meet the relevant stability conditions. Setting variables' desired values requires observing two items in the practical world: "trends in technology related to housing" and "trends in systems related to housing design." Observing trends in technology is significant for determining variables' desired values of material elements. Meanwhile, observing trends in systems related to housing design is necessary and useful for setting the desired values of most variables [7]. In this context, systems related to housing design include both compulsory and voluntary systems. Compulsory systems are building codes and regulations; voluntary systems are typically assessment and rating systems, standards, and design guidelines.

For example, if "durability" and "materials" have been identified as variables of "framework," the desired values of these variables can be determined in the following way. As for "durability," the system designers can set its desired value, considering a necessary level of framework's lifespan or deterioration resistance. Similarly, they can determine the desired value of "materials," based on a necessary level of utilizing materials which promotes resource-saving or waste-prevention, such as renewable, recycled, and recyclable materials. When determining the desired values of "durability" and "materials," it is significant to refer to relevant information in voluntary systems. Information on these desired values is usually included in voluntary systems, such as assessment and rating systems, and standards, whereas such information is outside the scope of building codes. Accordingly, system designers can set these desired values, by utilizing related criteria and information which are included in the relevant voluntary systems [7].

Meanwhile, if "resistance to earthquakes" has been identified as a variable of "framework," the desired value is a sufficient level of preventing damage caused by earthquakes. In quake-prone countries, quakeproofing standards are usually stipulated in the building codes. However, if system designers consider that the standard value specified in the building codes is insufficient, they make an addition to the standard value, so as to suit the desired value [7]. On the other hand, if they consider that the standard value is suitable to the desired value, they can use it as it is. In the latter case, the variable and its desired value can be omitted from the guidelines and checklist, because people naturally conform to compulsory building codes [7].

3.2 Process of revising the design guidelines and checklist

The "sustainable design guidelines" and "sustainability checklist" need to be revised, for adjustment to changing situations as well as for higher accuracy and user-friendliness. The revision process can be divided into three spheres:

(1) changes in the theoretical world, (2) changes in the practical world, and (3) feedback from the users. After making preparations from the above three perspectives, system designers modify the tables of the guidelines and checklist.

3.2.1 Changes in the theoretical world

Noticeable changes over time in the theoretical world are necessary to be reflected in the guidelines and checklist. First, after searching for recent changes in problems which affect the six stability conditions shown in **Figure 2**, system designers can amend the list of problems. Based on the amended list of problems, the system designers can also amend the list of the requirements for sustainable housing design. When amending these two lists, it is also necessary to observe the latest trends in "understanding about problems related to housing" and "understanding about sustainable housing." After that, system designers consider modifications on the "element-variable-desired value" structure of the guidelines and checklist.

3.2.2 Changes in the practical world

In addition to changes in the theoretical world, changes in the practical world need to be reflected. Changes over time in the practical world include "changes in technology related to housing" and "changes in systems related to housing design."

In order to search for "changes in technology related to housing," system designers need to watch the housing industry and regularly observe products related to housing. Housing-related products are closely connected with material elements in the guidelines. Therefore, if there are remarkable technological changes in housing-related products, such changes can be smoothly taken up in the guidelines, by arranging relevant material elements' variables and their desired values.

Meanwhile, changes in existing systems related to housing design also occur. Such changes include revisions or alterations of compulsory systems, including building codes, and voluntary systems, such as assessment and rating systems, and standards. System designers can include such changes in the guidelines, usually by modifying relevant variables' desired values.

3.2.3 Feedback from the users

As demonstrated in the lower part of **Figure 3**, "feedback from the system users" and "feedback from the home residents" are also necessary to be considered. The feedback from the system users is information about reactions to the guidelines and checklist, including comments on the validity and user-friendliness of these systems. Such information is used as a basis for the improvement of the systems. Meanwhile, the feedback from the home residents is information about reactions to the homes that have been designed with the guidelines and checklist. Such information, including comments on the homes' amenities and sustainability performance, is also useful for the improvement of the systems.

4. Illustration of producing and revising the design guidelines

4.1 Design guidelines produced in Japan

We produced sustainable design guidelines, with a mind to utilization in Japan. The English translation of the design guidelines has been demonstrated in Table 2 in the monograph's Chapter 4, "Methodology of Applying Control Science to

Sustainable Housing Design." This section briefly reviews the process of producing the design guidelines, anew following the three-stage production process shown in Section 3.1 and **Figure 3**. In addition, the design guidelines shown in the monograph's Chapter 4 were through a revision; however, this section considers the design guidelines as the newly produced version, for the convenience of explanation.

4.1.1 Identification of problems related to housing

The process of producing the guidelines starts with identifying environmental, social, and economic problems related to housing. In this case, identified main problems are shown in the second column of **Table 1**. In this table, the identified problems are divided into two types: "global/general problems" and "local/particular problems." Global/general problems include global warming and climate change, depletion of natural resources, waste, and increase of medical and nursing care expenses due to aging population. On the other hand, main problems considered as local/particular are short lifespan of houses, poor indoor thermal performance, damage caused by earthquakes, and poor landscape. In addition, the boundary between the two types is not so distinguishable.

4.1.2 Identification of the requirements for sustainable housing design

After determining the problems related to housing, we identified the requirements for sustainable housing design. For instance, "global warming and climate change" require "energy saving," "use of renewable energy," and "conservation of green space," for sustainable housing design. Similarly, "poor indoor thermal performance" demands "improvement of indoor thermal performance." In addition, items shown in the right column of **Table 1** are relevant stability conditions.

4.1.3 Determination of elements, variables, and their desired values

Based on the method of selecting material and spatial elements of homes, as well as the requirements for sustainable housing design, first we determined the elements in the design guidelines. The total number of elements was 26, made up of 14 material elements and 12 spatial elements. The material elements were framework, exterior, thermal insulation, windows and doors, interior, bathtub, piping, water heater, appliances, lighting fixtures, equipment for harnessing natural energy, equipment for rainwater use, water-using equipment, and outdoor facilities [3]. The spatial elements were total floor, specified bedroom, areas relating to water use and hot-water supply, position and area of windows, toilet, bathroom, stairs, doorways, hallway, main access route to the entrance, slope, and garden area [3].

After selecting the elements, we determined the variables and their desired values. Choosing only one element, that is, "thermal insulation," the rest of this section shows the details of determining the variable and its desired value. First, we identified "thermal insulation performance" as the variable, considering two requirements, namely, "improvement of indoor thermal performance" and "energy saving," as well as the relevant stability conditions. Higher thermal insulation performance contributes to occupants' better "health" as well as "environmental preservation" and "sustainable use of natural resources" through a reduction in energy usage for heating and air conditioning.

After that, observing trends in systems and technology related to housing thermal insulation performance, we set the desired value of the variable. The set

Type of problems	Main environmental, social, and economic problems related to housing	Requirements for sustainable housing design	Stability conditions
Global/general problems	• Global warming and climate change	• Energy saving • Use of renewable energy • Conservation of green space	• Enviro-preservation • Sustainable resources
	• Depletion of natural resources • Waste	• Extension of housing lifespan • Use of resource-saving or waste-prevention materials	• Sustainable resources
	• Harmful influences caused by climate change	• Adaptation measures	• Health • Safety
	• Flood risks due to rainwater flowing out • Water shortage risks	• Rainwater permeation into the ground • Water saving • Use of rainwater	• Enviro-preservation • Sustainable resources • Health • Safety
	• Increase of medical and nursing care expenses due to aging population	• Accessible and universal design	• Health • Safety
Local/particular problems (in Japan)	• Short lifespan of houses	• Extension of housing lifespan	• Enviro-preservation • Sustainable resources
	• Poor indoor thermal performance	• Improvement of indoor thermal performance	• Health • Enviro-preservation • Sustainable resources
	• Damage caused by earthquakes	• Higher resistance to earthquakes	• Safety
	• Poor landscape	• Consideration for landscape	• Health

Table 1.
Problems related to housing and requirements for sustainable housing design identified for the design guidelines produced in Japan.

desired value was "Grade 4" in the "energy-saving action grades (thermal insulation performance grades)" of the Japan Housing Performance Indication Standards (JHPIS) [3]. "Grade 4" is the highest in the above grades. In addition, housing thermal insulation performance is not stipulated in the building codes in Japan. Accordingly, utilizing the JHPIS, a national voluntary system, we determined the desired value.

4.2 The latest revision of the design guidelines

The design guidelines, the production process of which has been illustrated in the above, have most recently been revised. This latest revision has dealt with the three perspectives previously mentioned: (1) changes in the theoretical world, (2) changes in the practical world, and (3) feedback from the users.

4.2.1 Changes in the theoretical world

First, observing recent trends in understanding about housing-related problems, we have searched for the problems that affect conditions for stability. As a result, we have identified four additional problems which should be dealt with, as shown in the second column of **Table 2**. Two of them are global/general problems; the other two are local/particular problems. Based on these four problems, additional requirements for sustainable housing design have also been identified. After that, these additional requirements have been incorporated into the structure of "element-variable-desired value." The following describes the essentials of the identification and incorporation processes, by each requirement.

4.2.1.1 Storage of electricity

The Paris Agreement of 2015 has aimed to hold global temperatures "well below 2°C above pre-industrial levels and to pursue efforts to limit the temperature increase to 1.5°C" [8]. According to the latest IPCC report of 2018, limiting global warming to 1.5°C compared with 2°C would reduce challenging impacts on ecosystems, human health, and well-being [9]. In such very recent situations, the use of renewable energy, especially solar and wind power generation, is rapidly growing in many countries [10]. However, the amount of electricity derived from solar and wind sources fluctuates with time of day, season, and random factors including weather. Accordingly, sharp increase in solar and wind power generation is also increasing breakdown risks in electricity systems [11, 12]. Responding to such changing situations, we have added "storage of electricity" as a requirement for sustainable housing design. Battery electricity storage systems are developing fast, with falling costs and improving performance [13]. In addition, storing electricity leads to securing emergency power source, which is one of adaptation measures against climate change.

When "storage of electricity" is incorporated into the guidelines, "storage battery" has been added as a new material element. In the guidelines, this "storage battery" has been placed just after an existing material element, "equipment for harnessing natural energy," since storage batteries are often installed together with solar power generation systems. Subsequently, two variables of this new element, namely, "type" and "linkage," have been identified. The desired value of "type" has been identified as "stationary battery or electric vehicle battery," because both types are acceptable. Meanwhile, the desired value of "linkage" has been determined to be "interconnection with the home electrical system."

4.2.1.2 Considerations for homeworking, telecommuting, and lifelong learning

Recently housing has been becoming more important as a place of work. The development of information technology and spread of the Internet are facilitating home-based businesses. Meanwhile, an increasing number of companies have been adopting telecommuting [14, 15]. A great advantage of working at home is compatibility with childcare or nursing care. Moreover, working at home is also favorable for the environment because it can reduce energy for commuting.

Meanwhile, life longevity is increasing the necessity of lifelong learning. According to a best-selling book, *The 100-Year Life: Living and Working in an Age of Longevity*, an increasing number of people experience multiple careers, instead of a conventional single career, and inevitably continue to learn [16]. Now people not only recreate themselves in their free time but also need to "re-create" themselves at various places [16].

Type of problems	Environmental, social, and economic problems related to housing	Requirements for sustainable housing design	Stability conditions
Global/general problems	• Breakdown risks in electricity systems due to increasing solar and wind power generation	• Storage of electricit	• Sustainable resources • Health (in crises) • Safety (in crises)
	• Insufficient considerations for homeworking, telecommuting, and lifelong learning	• Considerations for homeworking, telecommuting, and lifelong learning	• Self-realization
Local/ particular problems (in Japan)	• Problems resulting from insufficient communication	• Floor planning suitable for good communication among residents	• Mutual help • Self-realization
	• Declining trend of food production and farming population	• Considerations for food production and agricultural learning at housing sites	• Self-realization

Table 2.
Additional problems and requirements for sustainable housing design identified for the latest revision of the design guidelines.

In the past, homes were not considered important as places of work and lifelong learning. Therefore, if attempting to start working or learning at home, people sometimes face difficulties due to a shortage of space and facilities. Taking such situations into account, we have added "considerations for homeworking, telecommuting, and lifelong learning" as a requirement for sustainable housing design.

When "considerations for homeworking, telecommuting, and lifelong learning" is incorporated into the guidelines, "area(s) for working and learning" has been added as a new spatial element. In addition, we consider that the "area for learning" is used by both adults and children. After that, we have identified two variables of this new element, namely, "place(s) in the home" and "equipment." The desired value of "place(s) in the home" has been determined to be "in or near the living/dining room and kitchen area." A reason for this setting is that "in or near the living/dining room and kitchen area" is convenient for combining working or learning with childcare or nursing care. Moreover, this layout is also expected to facilitate communication among residents, as shown in **Figure 4(a)**. Meanwhile, the desired value of "equipment" has been set at "table/desk and shelf (fixed or movable) and Internet connection."

4.2.1.3 Floor planning suitable for good communication among residents

In Japan, many studies on sociology and housing have pointed out that inappropriate room planning is related to social problems, such as school nonattendance and social withdrawal [17–20]. In the Japanese housing market, there are many homes, the floor plans of which are like the conceptual layout shown in **Figure 4(b)**. In homes with such floor plans, children easily enter their private rooms without seeing other family members and can stay isolated. As a result, lack of communication can cause underdeveloped communication skills and moreover school nonattendance [17, 19]. School nonattendance and social withdrawal are closely related to domestic violence [19] and difficulty in entering higher-level schools and finding jobs. Therefore, aiming to prevent these social problems, we have identified "floor planning suitable for good communication among residents" as an additional requirement for sustainable housing design.

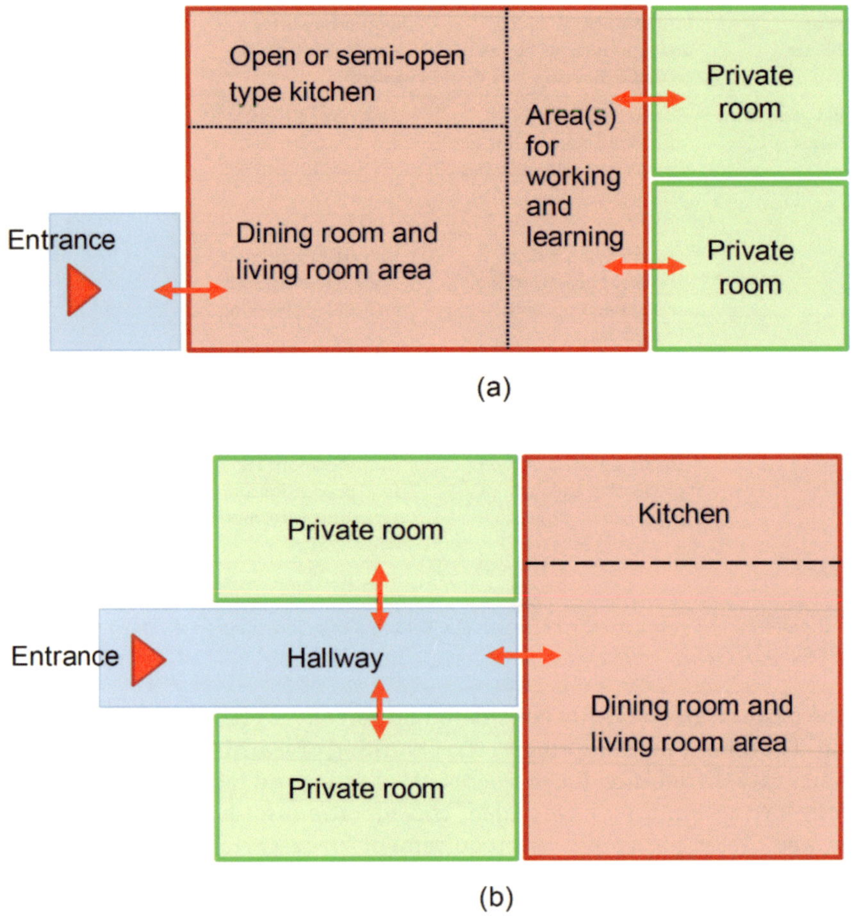

Figure 4.
Conceptual layouts favorable and unfavorable for communication.

The key space to communication among residents is the "living/dining room and kitchen area"; therefore, we have added this area to the design guidelines as a new spatial element. After that, we have identified two variables of this element, that is, "place in the home" and "type of kitchen." The desired value of "place in the home" has been determined to be "between the entrance and private room area," because many experts recommend this placement for more frequent communication [17–20]. Meanwhile, the desired value of "type of kitchen" has been set at "open or semi-open." Recently open- and semi-open-type kitchens are popular in the Japanese housing market since these types of kitchens are favorable for good communication, as compared with closed-type kitchens. The conceptual layout that includes these considerations, as well as "area(s) for working and learning," is demonstrated in **Figure 4(a)**. In addition, there has been a survey on more than 200 homes where successful examinees to famous junior high schools have lived. The results of this survey have revealed that most of these children have learned in or near living/dining room areas, while actively communicating with their family members [21].

4.2.1.4 Considerations for food production and agricultural learning at housing sites

The recent situation of Japan's farming population and food production has been declining alarmingly. The farming population has decreased from 2.61 million in 2010 to 1.75 million in 2018 [22]. The average age of the farming population has risen from 65.8 years old in 2010 to 66.8 in 2018 [22]. The total area of abandoned farmland has increased from 123,000 ha in 1980 to 423,000 ha in 2015 [23]. There has also been a long-term declining trend in the food self-sufficiency rate. The calorie-based food self-sufficiency ratio in 2017 is only 38% and the monetary-value-based self-sufficiency ratio is 65% [24].

Currently homes in Japan are usually separated from food production. One of the measures for encouraging people to be involved in food production is "food and agriculture education," which can be conducted at homes, schools, and local communities. Food and agriculture education at home is mainly connected with farmwork at vegetable gardens. According to a survey, experience of growing vegetables at home has positive effects on children's intention to engage in farming as work in the future [25]. Moreover, after starting as a vegetable gardener, more than a few people try to be a professional farmer [26].

Discovering that housing is related to Japan's food and agriculture problems, we have added "considerations for food production and agricultural learning at housing sites" as a requirement for sustainable housing design. In addition, there are also other reasons why this consideration has been added: (1) farm work is beneficial to health [27], (2) farming leads to securing emergency food, (3) farming encourages communication with family members and neighboring residents [27], and (4) farming can help people commune with nature.

When "considerations for food production and agricultural learning at housing sites" is incorporated into the guidelines, "vegetable garden and/or fruit trees" has been added as a new variable of the existing spatial element "garden area." Subsequently, the desired value of "vegetable garden and/or fruit trees" has been set at "included."

4.2.2 Changes in the practical world

Observing recent changes in technology and systems, we have revised the desired values related to two points: (1) thermal insulation performance standards, and (2) energy-efficient water heaters. Moreover, we have modified the desired value of rainwater equipment, following a slight change in the description of the related system.

4.2.2.1 Thermal insulation performance standards

Japanese housing thermal performance has traditionally been low. However, it is gradually improving, due to changes in housing-related technology as well as requirements for energy saving and occupants' health. As a result, recently a new national voluntary system, the "net zero energy house (ZEH) certification standards," has emerged and shown higher thermal performance criteria than the usual criteria [28, 29].

Recognizing these changes in the practical world, we have lifted the desired value regarding thermal performance in the guidelines. To be concrete, we have revised the desired value of "thermal insulation performance" of the two material elements, "thermal insulation" and "windows and doors," to the relevant criteria stipulated in the ZEH certification standards. The ZEH thermal insulation performance criteria are evaluated based on "average heat transmission coefficient (U_A)." The standard value of "U_A" varies, depending on climate classification. For instance, the criterion of "U_A" for the area where Tokyo is included has been set at "0.6 $W/(m^2 K)$ or less" [28, 29].

4.2.2.2 Energy-efficient water heaters

Following the rise of a new type of energy-efficient water heater in the market, we have modified the desired value of "type of water heater." At the same time, this revised version has adopted a way of offering choices as energy-efficient types, instead of quoting an assessment level used in Japan's national assessment and certification system, *CASBEE for Detached Houses*. The offered choices are the following four types: (1) solar, (2) electric heat-pump, (3) electric heat-pump and instantaneous gas combined, and (4) fuel-burning latent-heat recovery instant-supply. The third type, which is often called "hybrid," is the aforementioned new type of energy-efficient water heater [30].

4.2.3 Feedback from the users

4.2.3.1 Feedback from the system users

After finding a thought-provoking suggestion in recent feedback from the guidelines/checklist users, we have decided to include it in the latest revision. The main point of this suggestion is that lighting fixtures used in living spaces should be products with brightness and color adjustment functions, for energy saving and occupants' health.

Required brightness of indoor artificial lighting varies according to circumstances, such as natural lighting through windows and occupants' visual comfort. Moreover, lighting capable of adjusting brightness and color is also favorable for residents' health. Exposure to bright lights at night suppresses the secretion of melatonin and can affect sleep and potentially cause diseases, such as sleep disorder and diabetes [31, 32]. Although light of any color can suppress melatonin secretion, especially blue light at night has greater effects [31, 32]. Therefore, especially in living spaces, lighting fixtures fitted with brightness and color adjustment functions are favorable for both energy conservation and residents' health.

On the other hand, LED lighting fixtures with brightness and color adjustment functions are on the market at reasonable prices. Accordingly, we have made a revision to the desired value of "type of light" for "lighting fixtures" in the guidelines, namely "LED," adding "lighting fixtures used in the living spaces are fitted with brightness and color adjustment functions" as a supplementary note to "LED."

4.2.3.2 Feedback from the home residents

There have not been any reactions from the home residents which have led to the revision of the guidelines. Meanwhile, we have had a remarkable comment from the occupants who live in the home of the case study that has been introduced in Chapter 5 of the monograph. The comment says that "this home's thermal insulation performance is satisfactory, but this thermal performance level is never excessive." The building envelope's heat loss coefficient (Q) of this house is "Q = 1.9 [W/(m^2 K)]" [33], which can be converted into "U_A = 0.61 [W/(m^2 K)]." "U_A = 0.61 [W/(m^2 K)]" is almost equal to the net zero energy house (ZEH) thermal performance criterion for this area division. Accordingly, this comment has supported the adoption of the ZEH thermal performance criteria, as the desired value of thermal insulation performance.

Finally, all of the above revision items have been incorporated into the table of "element-variable-desired value" structure. The final revised version of the guidelines is shown in **Table 3**; the added and modified descriptions are written in *italics*.

Element	Variable	Desired value	Stability condition
Framework	Resistance to earthquakes	JHPIS 1-1: Grade 2 or over	• Safety
	Durability	JHPIS 3.1: Grade 3	• Sustainable resources
	Materials	CASBEE LR_H2 1.1: Level 4 or over	• Sustainable resources
Exterior (outer wall, roof, etc.)	Fire resistance (outer wall)	JHPIS 2-6: Grade 3 or over	• Safety
	Shape and color	Consideration for the landscape	• Health
	Durability	CASBEE Q_H2 1.2 and 1.3: Level 4 or over	• Sustainable resources
	Materials	CASBEE LR_H2 1.3: Level 4 or over	• Sustainable resources
Thermal insulation	Thermal insulation performance	*Thermal performance criteria stipulated in the net zero energy house (ZEH) certification*	• Health • Enviro-preservation • Sustainable resources
Windows and doors	Thermal insulation performance	*Thermal performance criteria stipulated in the ZEH certification*	• Health • Enviro-preservation • Sustainable resources
	Sunlight adjustment capability	CASBEE Q_H1 1.1.2: Level 4 or over	• Health • Enviro-preservation • Sustainable resources
	Sound insulation performance	CASBEE Q_H1 4: Level 4 or over	• Health
	Measures to prevent intrusions	CASBEE Q_H1 2.3: Level 4 or over	• Safety
	Protection of glass against impacts	With shutters	• Safety
Interior	Measures against formaldehyde	CASBEE Q_H1 2.1: Level 5	• Health
	Materials	CASBEE LR_H2 1.4: Level 4 or over	• Sustainable resources
Bathtub	Heat insulation	Insulated	• Enviro-preservation
Piping	Measures for maintenance	JHPIS 4.1: Grade 3	• Sustainable resources
	Method of water and hot-water piping	Header and pipe-in-pipe system	• Enviro-preservation • Sustainable resources
Water heater	Type of water heater	*Energy-saving type (solar, electric heat-pump, electric heat-pump and instantaneous gas combined, fuel-burning latent-heat recovery instant-supply type)*	• Enviro-preservation • Sustainable resources
Appliances	Energy-saving standard achievement rate	100% or more (three or more stars)	• Enviro-preservation • Sustainable resources

Element	Variable	Desired value	Stability condition
Lighting fixtures	Type of light	LED *(lighting fixtures used in the living spaces are fitted with brightness and color adjustment functions)*	• Enviro-preservation • Sustainable resources
Equipment for harnessing natural energy	Harnessed natural energy	100% or more of the total energy usage	• Health (in crises) • Safety (in crises) • Enviro-preservation • Sustainable resources
Storage battery	*Type*	*Stationary battery or electric vehicle battery*	• Health (in crises) • Safety (in crises) • Sustainable resources
	Linkage	*Interconnection with the home electrical system*	
Equipment for rainwater use	Rainwater equipment	*Rainwater tank (80 lit. or more) or system with a rainwater tank (80 lit. or more) for daily use*	• Health (in crises) • Safety (in crises) • Enviro-preservation • Sustainable resources
Water-using equipment	Water-saving functions	CASBEE LR_H1 2.1: Level 4 or over	• Enviro-preservation • Sustainable resources
Outdoor facilities (fence, etc.)	Form	Not blocking sightlines	• Safety • Mutual help
	Appearance	Consideration for the landscape	• Health
	Materials	CASBEE LR_H2 1.5: Level 5	• Sustainable resources
Total floor	Total floor area	75 m^2 or more [Note 4]	• Health
Specified bedroom	Routes to toilet and bath area, dining room, kitchen, and entrance	Accessible without steps	• Health • Safety
	Internal floor space	9 m^2 or more	
Living/dining room and kitchen area	*Place in the home*	*Between the entrance and private room area*	• *Mutual help*
	Type of kitchen	*Open or semi-open*	
Area(s) for working and learning	*Place(s) in the home*	*In or near the living/dining room and kitchen area*	• *Mutual help* • *Self-realization*
	Equipment	*Table/desk and shelf (fixed or movable) and Internet connection*	
Areas relating to water use and hot-water supply	Areas in the home	Placing them closer	• Enviro-preservation • Sustainable resources
Position and area of windows	Natural ventilation	CASBEE Q_H1 1.2.1: Level 5	• Health • Enviro-preservation • Sustainable resources
	Ratio of total window area to floor area in each living space	20% or more	
Toilet	Internal length or spacing	JHPIS 9.1: Grade 3 or over	• Health • Safety
	Handrails which help users sit and stand	Installed	

Element	Variable	Desired value	Stability condition
Bathroom	Floor space and width	JHPIS 9.1: Grade 3 or over	• Health • Safety
	Handrails help users go in and out of the bathtub	Installed	
Stairs	Grade of steepness	JHPIS 9.1: Grade 3 or over	• Health • Safety
	Handrails	Installed	
Doorways	Differences in level	No differences	• Health • Safety
	Width	75 cm or more (Bath: 60 cm or more)	
Hallway	Width	78 cm or more	• Health • Safety
Main access route to the entrance	Surface	Level or sloping	• Health • Safety
	Width	90 cm or more	
Slope	Grade of steepness	1/8 or less	• Health • Safety
	Handrails	Installed	
Garden area	Ratio of the garden area to the exterior area	40% or more	• Enviro-preservation
	Vegetable garden and/or fruit trees	*Included*	• *Health* • *Mutual help* • *Self-realization*

(1) "Material elements" are from "framework" to "outdoor facilities," and "spatial elements" are from "total floor" to "garden area."
(2) JHPIS means the Japan Housing Performance Indication Standards (for new homes).
(3) CASBEE means CASBEE for Detached Houses (for new construction)—Technical Manual 2018 Edition.
(4) At least one story's area (excluding stairs) is 40 m^2 or more.

Table 3.
The latest revised version of the sustainable design guidelines.

5. Discussion

This section discusses the study results from the following three perspectives: (1) comprehensive visualization of process for producing and revising the guidelines, (2) adjustment to different and changing situations, and (3) meaning of using the guidelines and checklist.

5.1 Comprehensive visualization of process for producing and revising the guidelines

Section 3 of this chapter has demonstrated the process of producing and revising the sustainable design guidelines and sustainability checklist. Especially **Figure 3** has comprehensively visualized both the production and revision processes in one diagram.

The production process has been shown through the three stages: (1) identification of problems related to housing, (2) identification of the requirements for sustainable housing design, and (3) determination of elements, variables, and their desired values. Following this three-stage process, Section 4.1 has briefly reviewed the production process of the already produced design guidelines. The previous publications, including the monograph, mentioned the production process and explained especially the third stage in detail. However, they could not show the consecutiveness of the three stages. On the other hand, the diagramming of the production process has clarified the overall perspective of producing the design guidelines.

Meanwhile, the revision process has been demonstrated from the three perspectives: (1) changes in the theoretical world, (2) changes in the practical world, and (3) feedback from the users. Following this process, we have made the latest revision of the design guidelines, as shown in Section 4.2. When comparing this latest revision with the previous one, we consider that this schematization has made the process more manageable and helped us to revise the guidelines more efficiently.

5.2 Adjustment to different and changing situations

The previous publications have already indicated that the "element-variable-desired value" structure in the guidelines has a mechanism of "adaptability to regional differences" and "flexibility toward changes over time" [4, 7]. This chapter has more concretely supported this indication. It has shown how to help system designers to adjust the guidelines to different and changing situations.

Tables 1 and **2** in Section 4 have demonstrated local/particular problems observed in Japan, in addition to global/general problems. As a result, such local/particular problems have naturally been taken in the process of producing and revising the guidelines. The produced and revised guidelines are suitable for the Japan's situation. If the same method is adopted in other countries or regions apart from Japan, similar results are expected to be obtained.

Meanwhile, Sections 3.2 and 4.2 have shown the process of revising the guidelines and its concrete example, respectively. These study results include theoretical and practical ways to adjust the guidelines to changing situations over time.

5.3 Meaning of using the guidelines and checklist

The guidelines and checklist are "user-friendly" because of their compactness as well as ease of comparison. The guidelines and checklist are simple tables. Meanwhile, material and spatial elements are equivalent to actual parts of homes; therefore, people involved in design can easily compare them with the actual home or drawings [4, 7].

The guidelines and checklist have another characteristic, "comprehensiveness." The main factor of this feature is the use of the "model of sustainability," which has been shown in **Figure 2**. This model contains "mutual help" and "self-realization," in addition to common considerations, such as health, safety, environmental preservation, and natural resources. As a result, as demonstrated in the latest revision of the guidelines, considerations for communication, working, and learning have been included. Furthermore, these inclusions have led to indicate a new aspect of homes, where residents are learning and working while communicating with one another.

The sustainable design guidelines and sustainability checklist can be used in various steps in the design processes, as demonstrated in **Figure 1.** If system users refer to them in such processes, they can easily and comprehensively check points for sustainable housing design. Houses are used for a very long time. Meanwhile, renovation after the construction wastes resources, labor, and money. In order to achieve sustainable housing as well as avoid regrets after the construction, it is meaningful to use the guidelines and checklist.

6. Conclusion

The previous study has provided the control system for promoting sustainable housing design where the sustainable design guidelines and sustainability checklist are incorporated. Based on these accomplished research results, this study has comprehensively visualized the process of producing and revising the design guidelines and checklist. Following this process, this study has also illustrated the production and revision processes of the design guidelines. The study results suggest that the comprehensive visualization can make these processes more manageable and help system designers to produce and revise the guidelines more efficiently.

Moreover, this study has indicated how to adjust the guidelines to different and changing situations. It has included a method of identifying housing-related problems, in which local/particular problems are identified, in addition to global/general problems. If this method is adopted in a country or region, the produced and revised guidelines are expected to suit to the situation of that country or region. Meanwhile, the provision of the process of revising the guidelines and its concrete example helps system designers to adjust the guidelines to changing situations over time.

The guidelines and checklist have a characteristic of comprehensiveness, resulting from using the model of sustainability. This model contains mutual help and self-realization, in addition to usual considerations, such as health, safety, and environmental preservation. As a result, the latest revision of the guidelines has led to include considerations for communication among residents, as well as working and learning at home.

Utilizing the "control system for promoting sustainable housing design" and the "model of sustainability," this study has illustrated the "process of producing and revising the design guidelines and checklist." These schematizations form a comprehensive strategy for promoting sustainable housing design. This comprehensive housing strategy is expected to be used to maximize people's well-being and minimize the environmental load.

Author details

Kazutoshi Fujihira
Institute of Environmentology, Inagi, Tokyo, Japan

*Address all correspondence to: fujihira@kankyogaku.com

References

[1] IPCC. Climate Change 2007: Synthesis Report. Contribution of Working Groups I, II and III to the Fourth Assessment Report of the Intergovernmental Panel on Climate Change. Geneva, Switzerland: IPCC; 2007. p. 104

[2] IPCC. Climate Change 2014: Synthesis Report. Contribution of Working Groups I, II and III to the Fifth Assessment Report of the Intergovernmental Panel on Climate Change. Geneva, Switzerland: IPCC; 2014. p. 151

[3] Fujihira K. Methodology of applying control science to sustainable housing design. In: Fujihira K, editor. Sustainable Home Design by Applying Control Science. Rijeka, Croatia: IntechOpen; 2017. DOI: 10.5772/intechopen.71324. Available from: https://www.intechopen.com/books/sustainable-home-design-by-applying-control-science/methodology-of-applying-control-science-to-sustainable-housing-design

[4] Fujihira K. System control for sustainability: Application to building design. In: Thomas C, editor. Complex Systems, Sustainability and Innovation. Rijeka, Croatia: IntechOpen; 2016. DOI: 10.5772/65875. Available from: https://www.intechopen.com/books/complex-systems-sustainability-and-innovation/system-control-for-sustainability-application-to-building-design

[5] Denton F, Wilbanks T, Abeysinghe A, Burton I, Gao Q, Lemos M, et al. Climate-resilient pathways: Adaptation, mitigation, and sustainable development. In: Field C et al., editors. Climate Change 2014: Impacts, Adaptation, and Vulnerability. Part A: Global and Sectoral Aspects. Contribution of Working Group II to the Fifth Assessment Report of the Intergovernmental Panel on Climate Change. Cambridge, United Kingdom/New York, NY, USA: Cambridge University Press; 2014. pp. 1101-1131

[6] Fujihira K. Basic schemes: Preparations for applying control science to sustainable design. In: Fujihira K, editor. Sustainable Home Design by Applying Control Science. Rijeka, Croatia: IntechOpen; 2017. DOI: 10.5772/intechopen.71325. Available from: https://www.intechopen.com/books/sustainable-home-design-by-applying-control-science/basic-schemes-preparations-for-applying-control-science-to-sustainable-design

[7] Fujihira K. Discussion and conclusion: Effectiveness, characteristics and future prospects of the methodology. In: Fujihira K, editor. Sustainable Home Design by Applying Control Science. Rijeka, Croatia: IntechOpen; 2017. DOI: 10.5772/intechopen.71321. Available from: https://www.intechopen.com/books/sustainable-home-design-by-applying-control-science/discussion-and-conclusion-effectiveness-characteristics-and-future-prospects-of-the-methodology

[8] Climate Focus. The Paris Agreement Summary [Internet]. 2015. Available from: https://climatefocus.com/sites/default/files/20151228%20COP%2021%20briefing%20FIN.pdf [Accessed: January 21, 2019]

[9] IPCC. Global Warming of 1.5°C [Internet]. 2018. Available from: https://www.ipcc.ch/sr15/ [Accessed: January 21, 2019]

[10] International Energy Agency. Key Renewables Trends 2016 Excerpt from: Renewables Information [Internet]. 2016. Available from: http://www.buildup.eu/en/practices/publications/

renewables-information-2016-key-renewables-trends-excerpt-0 [Accessed: January 21, 2019]

[11] U.S. Department of Energy. Grid Energy Storage [Internet]. 2013. Available from: https://www.energy.gov/sites/prod/files/2014/09/f18/Grid%20Energy%20Storage%20December%202013.pdf [Accessed: January 21, 2019]

[12] European Commission Directorate-General for Energy. DG ENER. Working Paper: The Future Role and Challenges of Energy Storage [Internet]. Available from: https://ec.europa.eu/energy/sites/ener/files/energy_storage.pdf [Accessed: January 21, 2019]

[13] International Renewable Energy Agency (IREA). Electricity Storage and Renewables: Costs and Markets to 2030 [Internet]. 2017. Available from: https://www.irena.org/publications/2017/Oct/Electricity-storage-and-renewables-costs-and-markets [Accessed: January 21, 2019]

[14] Global Workplace Analytics. Latest Telecommuting Statistics: Telecommuting Trend Data (Updated July 2018) [Internet]. 2018. Available from: https://globalworkplaceanalytics.com/telecommuting-statistics [Accessed: January 21, 2019]

[15] Suzuki H, Imaizumi C, Tachi Y. NTT Technical Review: Telework Trends [Internet]. 2012. Available from: https://ntt-review.jp/archive/ntttechnical.php?contents=ntr201203fa6.html [Accessed: January 21, 2019]

[16] Gratton L, Scott A. The 100-Year Life: Living and Working in an Age of Longevity. London: Bloomsbury Business; 2017. p. 407

[17] Goto N. Research on Family Relationships and Housing Floor Plans (*Kazoku-Kankei-to-Jutaku-no-Madori-no-Kenkyu*, in Japanese). JICE Report. Vol. 8. 2005. pp. 41-46

[18] Toyama T. Homes Where Familial Ties Are Strengthened (*Kazoku-no-Kizuna-wo-Tsukuru-Ie,* in Japanese). Tokyo: Heibonsha; 2007. p. 285

[19] Matsuda T. Building Houses and Destroying the Parent-Child Relationships (*Ie-wo-Tsukutte-Ko-wo-Ushinau*, in Japanese). Tokyo: Japan Housing Organization; 1998. p. 448

[20] Yokoyama A. Floor Planning Liable to Warp Children's Personality (*Kodomo-wo-Yugamaseru-Madori*, in Japanese). Tokyo: Joho Center Publishing; 2001. p. 258

[21] Shijima Y, Watanabe A. Homes where Children Grow Up Wisely (*Atama-no-Yoi-Ko-ga-Sodatsu-Ie*, in Japanese). Tokyo: Bungeishunju; 2010. p. 259

[22] Ministry of Agriculture, Forestry and Fisheries. Statistics on Agricultural Population (in Japanese) [Internet]. 2018. Available from: http://www.maff.go.jp/j/tokei/sihyo/data/08.html [Accessed: January 21, 2019]

[23] Ministry of Agriculture, Forestry and Fisheries. Devastated Farmland: Present Situation and Countermeasures (*Kohai-Noti-no-Genjo-to-Taisaku-ni-tuite, in Japanese*) [Internet]. 2017. Available from: http://www.maff.go.jp/j/nousin/tikei/houkiti/attach/pdf/index-4.pdf [Accessed: January 21, 2019]

[24] Ministry of Agriculture, Forestry and Fisheries. Japan's Food Self-Sufficiency Rate (in Japanese) [Internet]. 2018. Available from: http://www.maff.go.jp/j/zyukyu/zikyu_ritu/012.html [Accessed: January 21, 2019]

[25] Oura Y, Yamada I, Kataoka M, Yamamoto J. Observing the effects of school luncheon and food agriculture education on children (in Japanese). Journal of Rural Problems. 2009;**45**:254-257. DOI: 10.7310/arfe.45.254

[26] Japan Agricultural Cooperatives (JA). What is Food and Agriculture

Education? (in Japanese) [Internet]. 2019. Available from: https://life.ja-group.jp/education/description/ [Accessed: January 21, 2019]

[27] NTT Data Institute of Management Consulting. Survey Report on Methods of Grasping Evidence of Relationship between Farm Work and Health (in Japanese) [Internet]. 2013. Available from: http://www.maff.go.jp/j/study/syoku_vision/kenko/pdf/houkoku.pdf [Accessed: January 21, 2019]

[28] Energy Efficiency and Conservation Division Agency for Natural Resources and Energy Ministry of Economy, Trade and Industry. Investigation Results of the ZEH Roadmap Examination Committee (in Japanese). 2015. Available from: http://www.meti.go.jp/press/2015/12/20151217003/20151217003-1.pdf [Accessed: January 21, 2019]

[29] Energy Efficiency and Conservation Division Agency for Natural Resources and Energy Ministry of Economy, Trade and Industry. Definition of ZEH and Future Measures Proposed by the ZEH Roadmap Examination Committee [Internet]. 2015. Available from: http://www.enecho.meti.go.jp/category/saving_and_new/saving/zeh_report/pdf/report_160212_en.pdf [Accessed: January 21, 2019]

[30] Mae M. Household Energy Efficiency in Japan: Efficiency of Hot Water Supply Systems [Internet]. 2013. Available from: https://aceee.org/files/pdf/conferences/hwf/2013/3B-mae.pdf [Accessed: January 21, 2019]

[31] Harvard Health Publishing. Blue Light has a Dark Side: What is Blue Light? The Effect Blue Light has on Your Sleep and More [Internet]. 2012-2018. Available from: https://www.health.harvard.edu/staying-healthy/blue-light-has-a-dark-side [Accessed: January 19, 2019]

[32] Ayaki M, Morita T, Tsubota K. Biological Effects of Blue Light Contained in Artificial Lighting on Circadian Clock and Sleep/Awake Cycle [Internet]. 2015. Available from: https://www.jstage.jst.go.jp/article/jusokenronbun/42/0/42_1408/_pdf/-char/ja [Accessed: January 21, 2019]

[33] Fujihira K. Case study: Detached house designed by following the control system. In: Fujihira K, editor. Sustainable Home Design by Applying Control Science. Rijeka, Croatia: IntechOpen; 2017. DOI: 10.5772/intechopen.71323. Available from: https://www.intechopen.com/books/sustainable-home-design-by-applying-control-science/case-study-detached-house-designed-by-following-the-control-system

Chapter 5

Social Innovation and Environmental Sustainability in Social Housing Policies: Learning from Two Experimental Case Studies in Italy

Rossana Galdini and Silvia Lucciarini

Abstract

This chapter critically examines approaches and solutions developed by social housing to sustainably respond to the housing emergency plaguing contemporary cities and Italian cities in particular. In a broader perspective, we also investigate how housing has become 'difficult' in Europe and the poorest segments of the population run the risk of having their right to housing dramatically denied. Analysing housing in terms of its procedural dimension, we focus on two Italian case studies that evoke a new way of inhabiting the city, cases in which high standards characterised social housing and yet remain accessible to all. The Sharing hotel residence in Turin and Zoia social housing in Milan combine housing with other socially innovative measures in a framework of sustainability and avant-garde construction. These are significant examples that speak to issues such as temporariness, flexibility and the coordination of measures. These two cases both pursued objectives having to do with social, planning, architectural and environmental quality, albeit each in their own way. There are by now numerous examples of social housing in Europe and these have recently attracted growing interest in Italy as well; in this country, however, such projects represent valid instances of experimentation but are not at all widespread.

Keywords: sustainability, social innovation private-public housing policies, sharing and temporary use

1. Introduction

More and more, the image, economic logics and functions of contemporary cities reflect today's globalised society. In recent decades, however, urban designers have often produced architectural forms that are standardised and unresponsive to their context [1]. This self-referential type of architecture enjoyed success in that it met the demands of spectacularization and market logics, but it also contributed to undermining other fundamental aspects such as the representative character of a local place, community or era.

In opposition to these trends stands the concept of social architecture as a means of engaging with global phenomena, a discipline that takes cultural and social issues into account and is more in touch with the changes in economic and social structures and environmental problems characterising the twenty-first century. Architecture understood as a 'collective endeavour' entails a significant participatory component and a project of this kind must enjoy the involvement of all its stakeholders if it is to be implemented effectively and give rise to a democratic space.

Today we have an evermore urgent need for the kind of tireless ethical, political and social commitment on the part of those dealing with the city and its spaces that authors such as De Carlo [2] suggested in the past.

Architects, urban planners and scholars from all over the world support the idea, as Aravena[1] suggested, of "an architecture" that brings about concrete improvements in the lives of its various users.

Planning, innovation and new models of governance, together with a more mindful use of natural resources, are harnessed to the task of providing concrete solutions to people's needs. This approach to urban planning linked to the concept of regeneration involves a variety of different professionals, interests and aspirations and the best place to observe it at work is the arena of housing policy.

While the issue of housing represents a priority by virtue of the importance a home holds in a person's life, well-being and identity, at the same time this area is currently undergoing a serious crisis, particularly in Italy. In response to new demographic and social scenarios, the housing issue must be revisited from the standpoint of political choices and concrete responses at the local level even while taking into account a global context that brings its own effects and impacts [3].

In this context, social housing represents a heterogeneous set of measures both public and private, involving the state as well as the market, measures that consider spatial aspects in terms of their social implications and develop a variety of responses to different needs.

Social housing includes and supplements different areas: urban planning as well as architectural, economic-financial and social spheres. As a set of innovative policies for housing people, examples of social housing test out new solutions for contemporary living, paying particular attention to social, economic and environmental aspects.

The usual types of buildings we are accustomed to constructing do not always meet the housing needs of today's multifaceted society. As a result, practices of self-construction, forms of shared living, social condominiums and temporary housing are becoming more and more common in contemporary cities.

The new, widespread demand for housing pushes us to move beyond traditional patterns and find alternative solutions. And there is an increasingly wide and diversified public expressing this demand, from young people and non-resident students and workers to single-parent families, elderly people and migrants. Recently, in addition to the need for more housing, public administrations have also enlarged their urban public agendas to include measures for fostering social innovation and the ecological sustainability of buildings. In particular, the growing awareness regarding environmental issues in this area has led policymakers to adopt policies, methods and tools that focus specifically on natural and cultural habitats. There is a growing demand for spatial initiatives aimed at improving usability and

[1] Chilean architect Alejandro Aravena was the artistic director of the 15th Venice Architecture Biennale. His commitment in the field of social housing allowed him to win the 2016 Pritzker Prize and his designs are particularly focused on publicly-oriented projects with a strong social impact, committed to combining the needs of less privileged users with a focus on sustainability.

accessibility as well as making use of new materials. To apply the theme of sustainability to housing requires a set of actions that bring together spatial aspects, the social sphere, attention to resources and the overall quality of living spaces. When successful, these innovative solutions are characterised by a high degree of social engagement and give rise to virtuous practices of cooperation and sharing.

One example of this trend is the growth of non-profit organisations such as Architects Without Borders (AWB) operating in the countries of the Global South. Specifically, AWB aims to use architectural tools in the service of international cooperation. As Camillo Magni [4] notes, 'since 1998 they aim to facilitate processes of equitable and sustainable development through actions in the field of socially responsive design. They pay special attention to the investigation and spreading of appropriate and appropriable building technologies, as tools that can enable community participation and contribute to the emancipation of people in the production of their own habitat' [4]. We are seeing more and more projects designed to offer new opportunities to underprivileged people and designed to effectively face any critical issues that might arise while paying attention to energy consumption and environmental, economic and social sustainability.

In light of these considerations, in this chapter, we critically examine the approach and solutions developed by social housing to sustainably respond to the housing emergency plaguing contemporary cities and Italian cities in particular. In a broader perspective, we also explore how housing has become 'difficult' in Europe and the poorest segments of the population run the risk of having their right to housing dramatically denied.

Analysing housing in terms of its procedural dimension, we focus on two Italian case studies that evoke a new way of inhabiting the city, cases in which social housing is characterised by high standards and yet remains accessible to all. The Sharing hotel residence in Turin and Zoia social housing in Milan combine housing with other socially innovative measures in a framework of sustainability and avant-garde construction. These are significant examples that speak to issues such as temporariness, flexibility and the coordination of measures. The first project was developed to meet a temporary spike in housing demand on the part of vulnerable households; the second project aimed to build a liminal district in a site that has only recently been incorporated into the urban fabric of Milan. Both these two cases pursued objectives having to do with social, planning, architectural and environmental quality, albeit each in their own way.

There are by now numerous examples of social housing in Europe and these have recently attracted growing interest in Italy as well; in this country, however, such projects represent valid instances of experimentation but are not at all widespread. Nevertheless, the various disciplines dealing with this issue, as well as the current housing policy, clearly need to change the way they view this issue.

The recent debate on new social architecture proposes that we adopt a 'with people in mind' approach. Given their ethical and political engagement, the disciplines involved in this project prompt us to think about possible future directions. At the beginning of this century, De Carlo [2] suggested that we try turning the telescope the other way round, that is to say, modifying our customary way of reading the city. Indeed, housing and its policies require different lenses and, above all, the search for new focal points.

Considering housing policies from a sociological perspective, the focus was on the notion of social sustainability.

The content is divided in two parts. The first part of this study aimed at framing housing emergency and on the housing state-of-the-art in Europe where in the last 20 years a series of interesting measures were implemented. The primary objective of these programmes was to combine urban development with sustainable methods and approaches.

Building on these considerations, in the second part, this chapter analysed two cases of innovative housing projects in Italy, specifically the case studies of Milan and Turin, as a good practice of mixed policies: urban, socio-economic and environmental. In the conclusions, we evaluated factors that promoted these virtuous practices and the possibility to transfer these models in other contexts.

2. The housing state-of-the-art in Europe

Housing has become a strategic element of the welfare policies of many European countries as in recent years they have implemented a series of interesting measures in the effort to combine the issues of urban growth and residency policies with the objectives of sustainability. These programmes represent an initial response to the significant challenges of the present and coming decades, to the need to take suitable actions to both renovate existing buildings and build new ones. Many European countries, albeit with very different timeframes, instruments and outcomes, are dealing with a growing demand for housing. The housing situation has different characteristics depending on national contexts and local specificities, but we can nonetheless identify some common elements characterising the current housing challenge. For example, demand is highly differentiated, there is a close relationship between housing problems and the socio-economic context, and public spending in the sector is generally decreasing even while the role of the public sector is being gradually redefined, resulting in shifts in the forms and methods through which housing policies are implemented [5, 6].

The State of Housing in the EU 2017, the landmark biennial overview of Europe's Housing sector produced by the Housing Europe Observatory capitalist stresses the fact that housing has become 'difficult' for European citizens and the poorest segments of the population run the risk of being dramatically cut off from the possibility of finding a home. The report clearly shows that the challenge facing the housing construction sector has reached emergency proportions, especially in recent years. Housing is Europeans' highest expenditure. According to the 2018 Eurostat Report, 'some 11.1 % of the EU-28 population spent 40% or more of their household disposable income on housing'. Two years after the previous edition of the State of Housing Report, housing markets across the EU have begun to accelerate once again. The alarming fact is that, in most countries, house prices are rising faster than people's income. There is a direct link between growing global inequality and housing. The report also shows that the income gap between tenants and landlords is widening in a number of countries and that young people and migrants seeking to enter the housing market face ever-growing barriers. The gap between local areas is also dramatic in that finding suitable and affordable housing in places with good job opportunities is becoming increasingly difficult. In particular, large cities are facing structural housing shortages exacerbated by recent waves of migration. The political class has offered only limited responses and the housing problem will continue to represent a key challenge in the coming years [7].

As far as the Italian context is concerned, housing policies have always played a marginal role. In recent decades, first, the real estate speculation and, then, the economic crisis have contributed to the resurgence of housing problems, now affecting increasingly large swathes of the population. This critical situation, especially evident in large European urban areas, is caused by factors such as the high percentage of owner-occupied houses in the Mediterranean countries, in particular, the scarcity of social housing, and the diminution of available housing resources.[2] In Italy, the percentage of people living in houses they own (71.9%) still overshadows the percentage of those renting (14.8%) and 9.6% do not pay rent on the properties where they live. Only 3.7% pay reduced rental rates, of which 75% live in publicly owned housing. There are 7 million unoccupied houses or second homes (ibid). Contemporary transformations are causing a process of social segmentation and fragmentation centred evermore frequently on the family and also extending to housing. The traditional household comprising a mononuclear modal family is now accompanied by other forms: singles, the elderly, young couples, immigrants and students, all individuals who have different needs and desires. Moreover, some demographic processes, such as increasing immigration and the ageing of the population in particular, also exert new forms of pressure on the demand for housing. Housing shortages are now a growing trend: the continuous increase in property prices—and, to an even greater extent, actual rental rates—has not been matched by an equally significant increase in real incomes. At the same time, public administrations in many countries have become less and less involved, putting increasing trust in market forces to meet the housing demand. The current economic recession has further exacerbated the housing issue, making it even more critical for public and private institutions and non-profit entities jointly to intervene [8]. In this context, however, it is the 'grey segment' of the population that attracts the most: those who do not belong to the poorest sections of the population and yet are unable to access the housing market. By favouring these intermediate sectors of housing demand, housing policies and social housing in particular neglect the needs of the poorest individuals or subjects who are experiencing housing marginalization [9].

Moreover, despite the benefits that housing practices may generate for some segments of the population, they also lead to more negative outcomes. A highly evident example of this is the stigma associated with the areas that have hosted the most social housing buildings in recent decades—areas lacking adequate economic and social infrastructure, spaces for socialising, afflicted by widespread urban blight stemming from obsolete structures—and related problems of alienation, segregation and vandalism. In view of past achievements and new economic and social dynamics, it is clear that social housing measures must be integrated into a broader set of public policies. Another perverse effect of these policies is that they tend to downsize the role of public agencies, reducing them to mere facilitators: when policies seek to valorise the role of private actors, they create the regulatory and economic conditions for the private sphere to consolidate its position. These aspects hamber the right to a dignified dwelling on the part of families that do not have a house and/or that risk eviction or are living in conditions of degradation [10]. In Italy, the current housing emergency calls for solutions that combat poverty and promote support strategies at the local level [8]. In this field, architecture is considered a social tool for innovation and sustainability. Indeed, we believe that architecture, art and the social sciences can contribute a great deal to exploring new strategies, improving quality of life and engaging communities in connecting people and institutions. The case studies presented here are intended to demonstrate how scientific knowledge, art and technology can influence social issues in the arena of housing projects. Considering architecture not only as a spatial concept but also a social one, we can focus on the way we relate to the society around us and

[2] The housing situation in Europe displays varied characteristics: while in Italy and Spain owner-occupied housing prevails over rented housing, in Germany the market is mainly based on rentals (57% of housing is rentals). As far as the development of social housing is concerned, Germany and the Netherlands lead the ranking with more than 30% of housing managed through social leases. In Italy and Spain, the situation is very different, with 4% and 1% of total social rental housing, respectively.

find ways to create links with the landscape and human relations. As Haiek [11] suggests, 'problematics of contemporary cities do not lie in what is technically possible, but in what is socio-culturally desired'.

As these case studies seek to show, it is more important now than ever to expand the role of social housing and put programmes, including experimental ones, in place; to support the move in this direction, it makes sense to assess and compare different initiatives at the European level, fostering an exchange of expertise and know-how among countries.

3. Social innovation and environmental sustainability in social housing policy

Today, the housing demand takes shape in a context made problematic by the need to invent different approaches to design, alternatives to the usual schemes, approaches that might lead us to develop alternative solutions. What is needed are measures acting on space aimed at improving the usability and accessibility of built environments, encouraging energy saving and facilitating their management and, above all, solutions that foster socialisation practices in part by redefining living spaces. The demand for housing involves both material and immaterial relations with the other elements of the urban system. It is a demand for infrastructure and services but also for less hasty processes that involve policy recipients through collaboration and sharing. Recent housing policies focus on housing *quality*, a broad term comprising a wide range of issues 'not only to the dwelling itself but also to the broader surrounding residential area in which people live' [12]. These planning projects combine quantitative and qualitative aspects to develop satisfactory responses to a demand that is increasing in both size and complexity. The specific objectives vary in different countries: recovering real estate stock, regenerating local areas, energy savings, and social cohesion and participation—in part because integration is the basic principle of housing.[3] One of the most innovative features that deserve to be highlighted is the fact that social housing seeks to address not so much or not only the need for housing as the need for living, in the broad sense, as a field that includes social relations, services and the quality of the urban landscape.

In recent years, the need to combine housing policies with the concerns of urban growth and objectives of environmental, economic and social sustainability has given rise to a series of interesting social housing programmes located mainly in northern and central European countries. These diverse practices all have in common the attempt to meet specific basic requirements such as the morphological and spatial quality of buildings, energy savings, cost containment, experimentation with technological innovation and other elements that are crucial to a good quality of life. Moreover, that is not all. As some of these good practices show, an environmental focus not only promotes ecologically oriented lifestyles, it also brings with it an idea of social cohesion and aids in transcending the logic of exclusion that created ghetto neighbourhoods and widespread urban blight in the suburbs of the past. Housing policies should meet the needs (geographic, economic opportunities, materials, construction processes and cultural context) of a specific community or place. The sum of these aspects is synthetically expressed in the term social innovation, a concept that has been interpreted in varying ways. The concept of social innovation is essential in light of the widespread recognition that the state, market and often civil society organisations as well prove inadequate to deal effectively with social problems.

[3] Irer final report, Creation of study paths aimed at evolving residential policies and social housing.

DOI: http://dx.doi.org/10.5772/intechopen.86279

In her 2013 study, Czischke examines different definitions of social innovation. The Young 2010 Foundation/Social Innovation eXchange and BEPA definition, for example, highlights the fact that such innovation seeks to respond to social problems through: 'new ideas (products, services and models) that simultaneously meet social needs (more effectively than alternatives) and create new social relationships or collaborations. While "innovation" refers to the capacity to create and implement novel ideas that are proven to deliver value, "social" refers to the kind of value that innovation is expected to deliver: a value that is less concerned with profit and more with issues such as quality of life, solidarity and well-being. Social innovations are innovations that are social in both their ends and their means'.

The concept of social innovation in housing policies refers, for example, to the role and actions of the organisations tasked with providing and managing housing. They intervene in areas in which institutions, the market and civil society have shown themselves lacking, such as housing, environmental sustainability (e.g., the improvement of energy efficiency and practices of reuse), inclusion at the neighbourhood level, training and labour market inclusion for tenants, and self-help housing initiatives. The spread of the idea of social innovation has also elicited criticism from those who consider this idea nothing more than a superficial label or an attempt to obtain more funding [13]. The hope is, however, that social innovation entails a paradigm shift and promotes real empowerment for people. Furthermore, the idea of sustainable housing implies providing affordable housing: this concept includes solutions that help residents improve their quality of life and make their sense of community stronger. The concept of sustainability applied to housing policies implies three levels of action, focused on the environmental sphere, the social sphere and sustainability.[4]

Some social housing projects focus on achieving the ultimate objective of more sustainable development. These measures allow us to consider how such projects inspired by the theoretical principles described here can be implemented and how, at the same time, these projects of implementation might reveal a shift of perspective and method both in terms of conceiving new projects and throughout the process of carrying them out, from theory to practice and vice versa.

The project by Aravena that in 2003 was able to create houses for 100 disadvantaged families in Inique, a city in northern Chile is particularly innovative interesting. The architect used funding from a public programme, but the budget was only sufficient to purchase the land. In line with his approach, the solution he developed was the simplest one. The accommodations handed over to residents consisted of a series of semi-built houses that they could easily complete on their own. The project, carried out by Elemental Studio, was inspired by this principle of creating the essential elements, whatever the residents were not able to design and build on their own in keeping with high-quality standards, namely the load-bearing structure, roof, utility systems and restrooms with running water. Everything else? This, called the 'good half house' is an unbuilt space that can later be filled according to the resident's individual possibilities, needs and taste. In this case, the idea of 'elemental' is joined by the idea of 'incremental'. Guaranteeing only a percentage of the building in order to grant freedom and autonomy to its future residents is an economically strategic solution that also has a highly ecological logic

---

[4] The house is designed to reduce gas emissions, save water natural resources and reduce waste during the house's lifetime; "Social sustainability: the house is designed to prevent injuries through built-in safety features. 3. Economic sustainability: the house is designed to save money during construction and over the lifetime of the house. The project design is focused on bringing those aspects together. Reuse and flexibility are so important, the aspect of sustainability played a key role in the design". http://www1.indstate.edu/facilities/sustainability/

and social aims. The flexibility and reversibility of a structure is key to ensuring its sustainability, but also important for respecting individuals and their specificities. The structures in Inique were built after consulting with the people who would be living in them [14]. These practices of social architecture involve the community in implementing the design in order to ensure higher chances of success, creating a sense of ownership, social cohesion, and later even attention to the tasks of building maintenance.

This design is simple and low cost, but not easily feasible in every context. Nonetheless, the principles inspiring this project are valid and reproducible. Aravena's social housing thus represents a tool for including more underprivileged segments of the population in the credit system by offering access to bank loans. This project enjoyed a high degree of approval but has also attracted criticism: according to some concerned observers, these methods have the potential to turn social housing into a means of creating debt and thereby reproducing capitalist relations [15]. In Chile, however, the project proved successful.

Residential policies, and social housing in particular, seem to gravitate around the central idea of creating democratic tools to redistribute services and opportunities but also enlarging the scope of urban quality for collective well-being. From 'housing policies', therefore, the focus shifts to 'living policies'. The principles of interaction, collaboration, and the participation of future residents are fundamental for these policies along with the possibility of developing shared designs. As we will show here below, this idea of 'reversing' perspective to place people at the centre of the process rather than simply framing them as the end users of buildings or the beneficiaries of profit-oriented policies can also be seen in our selected case studies in Turin and Milan, analysed here in the light of the specificity of the Italian context.

4. Material and methods

The central aim of this study was to examine to what extent social innovation and environmental sustainability can coexist in social housing policies.

For this reason, the chapter critically investigated approaches and solutions developed by social housing as a response to the housing emergency in contemporary European and Italian cities. Besides meeting house needs, housing policies are expected to promote interaction, social cohesion, cooperation, civic engagement and economic development.

Social housing is generally considered as a promising model for urban development, and most empirical case studies in Europe demonstrate the potential to create active and diverse communities, promoting and maintaining sustainable living environments.

Building on these considerations, we analysed two cases of innovative housing projects in Italy, specifically the case studies of Milan and Turin, as a good practice of mixed policies: urban, social economic and environmental.

The Sharing hotel in Turin is the first temporary social housing initiative; the Zoia is an example of social and cooperative housing in Milan.

May these projects represent an opportunity for housing demands, giving at the same time responses both to people needs and aspiration for a better quality of life?

To answer this research question, we adopted a historical narrative perspective Taylor and Ramsey [35], due to catching the evolution of the phenomenon and its links to urban and socio-economic aspects. The narrative perspective was developed in the case studies, emblematic in terms of actors, spaces and organisation's mission.

The analysis of these experiences in Milan and Turin aimed to observe how the different dimension of sustainability (economic, social and environmental) and affordability that characterise the concept and the practices of social housing promote innovative and successful housing solutions. The starting point of our study is that 'sustainability is first about the people, how they make choices and the consequences of their choices' [16].

King [36] affirms that people play a central role in creating environments and institutions that support human well-being and encourage collaboration. Since well-being depends on economic, cultural, social, political and environmental aspects, an interdisciplinary vision is required to address the human dimension of sustainability [17].

For Jaberdeen [37], urban form (the physical structure of the city) is the result of different patterns (transports, water, global climate etc.) that are interconnected. They can influence social interaction and human behaviour.

A central issue of this study is the idea of social sustainability, related to economic activity, social relationships and ecological impact [18]. The concept is connected with three fundamental ideas: social exclusion, social capital and governance; they play a central role in creating sustainable communities.

As Colantonio [19] highlights, the term social sustainability has three different and interrelated components: normative (a long-term vision), strategic (a wide range of activities) and descriptive (what is). According to many authors, social sustainability lies in the intersection and implies policy interaction of environmental, social and economic issues and a long-term perspective [18]. These different aspects influence urban policies: participation, justice, democracy, social cohesion become imperative.

How can this imperative be applied to different contexts and how can they be included in housing policies?

These categories are listed in the 'Egan Wheel' [20], which can be adapted to evaluate housing policies.

We adopted and modified Egan wheel and Manzi's (et al.) adapted scheme to evaluate to what extent the selected projects of social housing in Italy

Social structures	Spatial structures
Promoting well-being and liveability	Provision of common open spaces
Supporting resident participation	Provision of shared spaces, playgrounds, community gardens, squares
Encouraging social cohesion and interaction	Provision of quality public and open spaces to promote events, meetings, markets
Integrating services	Provision of efficient transportation, car sharing, bikes service, services that respond to the communities' demand
Facilitating partnerships	Provision of mixed-use spaces to encourage the creation of local business, job opportunities
Ensuring safety, security	Provision of accessible and safe public areas
Promoting access to communication	Offering access to digital communication networks
Providing the condition for opportunities for all	Promoting shared and individual spaces that can be transformed into places

Sources: adapted from [20] by the authors.

Table 1.
Indicators for evaluating housing policies.

may promote through the interaction of social and spatial structures social sustainability.

For this reason, we employed and compared a set of eight indicators, using the Egan wheel scale to construct a conceptual framework. Our aim was to verify if the selected case studies could represent valid models to be transferred in other contexts. **Table 1** summarizes the identified indicators:

The conceptual framework used for our analysis took into consideration both social and spatial structures of these social housing experiences. We considered these dimensions as elements that define the context, as well as devices for framing and characterising the case studies. For this reason, our analysis was based on secondary data from different sources such as The State of Housing in the EU 2017, the biennial compass of Europe's Housing, The Egan Review, and official documents (plans, projects, photos etc.), ad hoc analyses and studies developed in Milan and Turin concerning innovative housing programmes.

Based on this material, we compared the selected case studies, giving a description of our results, highlighting reflections that are relevant to the topic under investigation.

Meeting the basic demands, offering a wide range of services, encouraging social cohesion and interaction seem to be the main goals of a valid social housing project.

5. Two case studies in Italy

As many classic authors of literature on the city [21, 22] have noted, urban agendas and the mainstream scientific-cultural sphere developed a pressing need to combine development and inclusion in order to curb the riots breaking out in North American cities in the 1980s and in Europe, particularly France, in the 2000s.

These instances of social upheavals were caused by deepening inequalities in economic and living conditions, inequality that was assuming an increasingly ethnic shape and was spatially delimited to certain disadvantaged areas.

While these moments of crisis produced a great proliferation of studies on the relationship between cities and immigration, some scholars have also focused on the need to strengthen the potential of existing urban spaces [23] in economic, social and environmental terms. These studies focus on the city's cultural sphere by consolidating material and immaterial capital, thanks in part to the role of 'creative' innovators who promoted quality development in the city, part of cognitive and professional networks that are local but globally integrated (such as in Florida, for instance).

The focus was on these two poles: on the one hand, social stability/cohesion in deprived or highly segregated areas but without considering an urban narrative of inclusion and development; on the other hand, the potential to innovate and position projects within qualified networks focused on innovation and creativity. In the European context, regeneration and renewal programmes have been put in place to change vulnerable areas by transforming them into points of strength within existing cities. These measures draw much of their justification and meaning from an approach focused on integrating spatial/environmental, economic and social cohesion aspects.

Beginning in the 1990s, these approaches and analyses have been central in a number of capitalist, urban Western contexts and have represented one of the most important and innovative lines of investment in urban areas, giving rise to a rich and complex frame of reference for studies on the city. These analyses considered spheres which deploy along three axes: horizontal, vertical and transversal.

The horizontal axis is summarised quite effectively by Donzelot [24], who distinguishes between policies aimed at the local area (place) and those for residents (people). The vertical axis, on the other hand, seeks to understand the matrix of programmes and projects, whether they are managed from above (i.e. by institutions, administrators), called top-down, or developed from below and subsequently institutionalised through the reverse process, called bottom-up. The transversal axis, which intersects all the points of convergence of the previous axes, focuses instead on governmental instruments and devices (i.e. agreements, participation) and the arena of the actors involved (public, private and mixed).

In recent years, there have been instances of innovation in all three axes thanks to a higher degree of hybridization between policies related to place and those focused on people; a widening in the scope of actors involved in regeneration projects—public, private and from the third sector—and a broader range of instruments and institutional devices through which these partnerships develop. Two significant cases of this trend in the Italian landscape are the Sharing Hotel in Turin and the Zoia district in Milan.

These two cases have a common matrix: they are both based on broad public-private partnerships led by the local administrations. In the second half of the 2000s, these local governments issued calls for tenders focused on dilapidated state-owned spaces that could be regenerated through projects that would bring these blighted spaces back into the urban fabric. The sites were a former industrial area on the outskirts of Turin and a former agricultural area incorporated into the Milan metropolitan area as the city expanded over time. Both of these sites were characterised by a state of 'abandonment' and left to sit on the economic and social margins. The idea of redevelopment underlying both projects seeks to respond to increasingly diverse housing needs, not only among vulnerable subjects but also among those with 'grey' profiles. These are individuals who find themselves suddenly vulnerable as their family situations become more fragile or their employment prospects become uncertain.

These two projects are mainly aimed at developing housing systems that combine residency with associational and neighbourhood life, guaranteeing access to urban opportunities even in areas that were previously blighted or disconnected from the city. Each of the projects also aims to achieve a specific objective: in the case of Turin, offering a temporary lease, given that this particular social need still goes unmet by the available housing supply. In the case of Milan, creating a neighbourhood for family-type settlement that would be accessible to residents who find themselves expelled from the extremely expensive housing market in more central areas of the city.

The selected case studies reveal analogies in methods and regeneration programmes, as **Table 2** shows.

The common objective of the two projects is to combine housing capable of responding more effectively to the changing needs and conditions of individuals and families with an environment rich in community opportunities, promoting cultural and economic activities in order to increase social capital. At the same

Structural dimension of urban regeneration	Milan case (regeneration of a marginal area)	Turin case (urban reuse strategy in the urban core)
Horizontal axis	Mixed 'place' and 'people' policies	
Vertical axis	Top-down programmes	
Transversal axis	Partnership: public-private and third sector	

Table 2.
Main axis of regeneration programme, strong homogeneity between the selected case studies.

time, it focuses the design's architectural canons on environmental and sustainable aspects in an effort to increase residents' awareness about environmentally 'virtuous' behaviours such as recycling, reuse and rational energy consumption.

5.1 The Turin Case

The Turin-based project 'Sharing Torino' involved founding a specific company, SHARING s.r.l., set up by a social cooperative called DOC. The component of this project consisting of apartments for the grey segment of society—students, young couples, precarious workers and mid-term commuters who are currently unable to access the 'normal rental market'—promises high social impact. The initiative originated from a public appeal by the City of Turin, which issued a call for tenders focused on the design and management of the city's first Social Hotel. Sharing Torino thus collaborates with the Turin municipal administration to host families experiencing emergency housing conditions and with other public and private bodies to implement projects of solidarity at the international level as well.

The hotel consists of two large, 9-storey-tall buildings, offering 58 rooms to be used as 3-star hotel rooms and 122 fully furnished residential units with induction cooktops and free Wi-Fi. The utilities are controlled through a home automation system. The rates and terms are different according to the length of time residents stay there, making it possible to offer housing at contained costs and thereby concretely contributing to solving the city's housing emergency and needs.

The facility also offers an intense social and cultural programme carried out in collaboration with the neighbourhood and local associations, intended to be used both by those staying in the hotel and residents of the neighbourhood that hosts the project. The 10,000-square-metre building, a former post office, was transformed into a residence for temporary social housing; after a year and a half of renovation work, today its exterior features a colourful façade reminiscent of a Mondrian painting.

Thanks to highly flexible commercial options, Sharing is able to meet the most diverse of housing demands. Residents can stay in a room or apartment for periods ranging from 1 day to a full year—although this maximum limit is only possible for those with special needs. There is the Housing Formula, offering rooms with kitchen for stays of at least 12 months; the Campus Formula, developed for not only students but also young professionals and researchers, for stays of at least 6 months; the Residence Formula, for stays of more than 14 nights; and the Hotel Formula, for shorter periods. A furnished studio apartment costs 190 euros per month; for a larger apartment, individuals can spend up to a maximum of 459 euros.

Finally, there are also rent-control options for those who meet certain requirements: people with an income lower than 12,000 euros per year, people with an income lower than 20,000 euros if disabled, separated with dependent children, or legal immigrants, people over 65, students, people participating in government assistance programmes for the relatives of individuals hospitalised in hospitals belonging to the national health system, and people participating in social care programmes. As a matter of fact, this project is also aimed at providing a testing ground for those looking to experience autonomous living before approaching the private market. For the City of Turin, it represents a cost-free solution to the housing emergency: thanks to a memorandum of understanding signed with the city involving 25 apartments set aside for the city to allocate; in the first year the facility hosted 319 people suffering from emergency housing conditions. However, it is not only the most vulnerable individuals who turn to the facility. The composition of the Sharing population is quite heterogeneous, including students (60%), non-resident workers (16%), young couples (6.6%) and tourists. This case attests to the growing scope of that grey area of housing seekers who, although not experiencing emergency housing conditions, face serious difficulties in finding places that meet their needs on the regular market, in this case, mainly due to the temporary nature of the accommodation.

5.2 The Milan case

Zoia is a district on the outskirts of Milan, located in a former rural village that ended up being absorbed by the Milan urban area a few decades ago. It is connected to the city centre by a historic metro stop, Bisceglie, and a brand new one, San Siro; Trenno Park and Cave Park mark the border between this neighbourhood and the countryside adjacent to Sud Park. The neighbourhood itself grew out of some of isolated fragments of late modernist cities that became areas of social hardship and precariousness.

In 2008, the Municipality of Milan put eight of its property areas, granted under surface rights, out to tender with the aim of fostering the construction of social housing. The lots subjected to tender were granted 90-year surface rights for the purpose of constructing residential buildings to be rented at reduced, moderate or fixed-maximum rates and/or with use rights, with prefixed sale prices. Unlike the previous call issues by the City of Milan in 2005, 'Abitare Milano, Nuovi spazi urbani per l'edilizia sociale' (Living Milan, New urban spaces for social housing) addressed to architects, this call involving eight lots targeted actors specific to this sector, such as cooperatives and companies with internal teams of designers or technical experts.

Two cooperatives were awarded the lot in Via F.lli Zoia, one called CCL Solidarnosc and another called Edificatrice Ferruccio Degradi. The project resulted in the construction of 3 buildings—2 for sale and the third to be rented at reduced and fixed-maximum; of 95 apartments, 44 are for rent and 16 of these are for public housing.

In addition, the cooperatives were able to access regional funding from Servizi abitativi a canone convenzionato (SACC—housing services at fixed-maximum rates) for the rent-controlled apartments, thus capping rent at 79 €/sqm per year, including parking, while the public housing apartments are co-funded by the region through a framework agreement between the Municipality of Milan and the Lombardy Region. The average rent for these apartments is about 15 €/sqm per year.

The complex also includes 'creative' offices or workshop spaces and, thanks to a public competition held by the social coop Noicoop, the cooperatives assigned these spaces to a luthier, a group of set designers, an art gallery and two architects, who were committed to organising a series of activities for the inhabitants. The rental area also includes a space for professional start-ups: about 200 square metres allocated at a moderate cost to workshops/laboratories for young creative professionals and craftspeople promoting activities with a positive social impact on the neighbourhood. The complex looks out onto a private courtyard for public use as well as a public square.

The fact that the complex includes housing for both sale and rent ensures the creation of a social mix, a result which was the primary objective of the call. In this case, social housing is sustainable thanks to the low cost of the lots released by the public administration and, in particular, the fact that the area set aside for rentals only cost 1 euro; the project is also made possible by funding from the Lombardy Region dedicated specifically to social housing.

Large companies, real estate investors and public entities are not the only actors involved in transforming the city; in this case, the cooperative sector was included among the set of actors playing a leading role. Residents' cooperatives have been active in this area for decades: they built substantial components of our cities and continue to do so through a new business model that positions the people, cooperative members, at the beginning and end of the production chain; these members are both financiers and recipients of the construction project. The specific cooperatives behind ZOIA have historically represented models, belonging to two parallel worlds but now collaborating directly.

The first type is that of 'undivided ownership', as exemplified by FerruccioDegradi, a model that stems from historical workers' mutual aid societies as channelled, in Italy, through Legacoop. In this case, the cooperatives are the owners of the properties, while the apartments are given to cooperative members at very low, advantageous rates. In general, this type of cooperative also deals with property management and, as an operational and asset guarantee often sets up systems in which the members themselves pool their savings in the form of 'risparmi sociali' or social savings schemes.

The second type is 'divided ownership', as in the case of the Solidarnosc cooperative, an offshoot of the organisation Consorzio Cooperative Lavoratori or Workers' cooperative consortium. This organisation also has its roots in labour organising and mutual aid but specifically in a Catholic framework and belongs to the umbrella organisation Confcooperative. In this case, the members become the owners of the apartments after having followed the entire planning process, which often lasts many years. Both of these cooperatives share a spirit of mutual aid, and place people and their well-being at the centre of the planning process. Members, who need a home, come into an existing cooperative or establish a new one and entrust their savings to it, thus enabling the implementation of the project. Once construction is completed, the apartments are allocated at prices much lower than the market average because they reflect the real distribution of costs, with no overhead. The cooperatives are committed to offering the inhabitants services and opportunities for socialising by setting up a common hall and shared laundries on the top floor of the rental building, organising social and community events and supporting micro-entrepreneurship through the collaborative workspaces overlooking the central courtyard.

5.3 Social and sustainable housing: comparing two logics in Milan and Turin

The comparison between these projects puts in evidence some differences regarding their logics of action, as **Table 3** summarizes.

Logics of action	Milan	Turin
Main targets	Lower middle class, expelled by the high-priced market of the city core who share the value of living in an affordable middle-class neighbourhood	Grey zones of individuals (mainly professionals) in need for temporary housing solutions
Social innovation	Focused on economic dimension, in particular the sharing or collaborative ones (co-working and fablabs)	Focused on social cohesion, fostering common spaces and collective practices inside the building
Environmental sustainability	An environmentally low-impact neighbourhood designed and built using sustainable materials and technology	Reuse strategy using eco-sustainable solutions; high energy efficiency, low environmental impact

Sources: author's elaboration.

Table 3.
Logics of action pervading social housing policies.

Goals	Design strategies (Milan)	Design strategies (Turin)
Bringing quality to the built and natural environment	Distinctive paths, nodes, edges, landmarks that help to construct a sense of identity to the community	Provision of a wide range of sizes and types of housing units, to accommodate the different family configurations
Creating a sense of community identity and belonging	Provision of social and cultural infrastructure for the community: public and green spaces, community spaces to the development of projects that encourage diversity and integration	Provision of shared spaces in the community to encourage social integration
Establishing an environmentally sensitive relationship between places and people	Incorporation of renewable energy systems and passive strategies to buildings and the urban infrastructure	Encouraging recycling and reuse
Improving economic advantages	Creating laboratories Sport and learning activities for children	Creating job opportunities Saving energies and shortage costs

Sources: adapted from [20] by the authors.

Table 4.
Goals and design strategies for sustainable housing.

These differences in the logics of actions affected the definition of goals in the two case studies. Adopting a selection of the items identified by Egan [20], we distinguished different design strategies implemented in the two contexts, as synthesised in **Table 4**.

Our analysis showed the local contexts influenced and led the way that social housing is defined. Even in programmes that have a similar structure (as **Table 1** has shown), policies are adopted in an arena of multiple actors with various perceptions that defined different design strategies to achieve the same goals.

The evaluation of these programmes has to take into account these differences and implement a meta-analysis able to better understand the links between policy, goals, strategies and their congruity.

6. Conclusion

Recent socio-economic and environmental dynamics have contributed to exacerbating the demand for housing to the point that it has reached emergency proportions, particularly in Italy. Today's difficult situation calls for a change in perspective, moving towards a type of architecture that reconnects with its socially engaged roots: an architecture built for the community, to improve quality of life for those who find themselves in situations of severe or temporary vulnerability.

The idea of social architecture links up with the need for housing policies that are 'more social'. Fortunately, projects in Europe seeking to reframe the right to housing by introducing an additional social component for the poor [9] show that this is currently a feasible objective. This new social offering is especially significant for people who have experienced conditions of exclusion and hardship, allowing them to access real housing opportunities. According to this vision, social housing represents a field of experimentation that is unquestionably effective in developing solutions that meet people's needs, projects capable of combining spatial design and economic considerations while respecting environmental constraints. Some social housing experiences contain innovation even in the methods and procedures.

The Turin case, for example, is the first example of temporary housing project. Temporariness favours small-scale and bottom-up initiative. Today, temporary use affords different actors to explore informal solutions, allow for flexible and transitory responses and simplify some bureaucratic planning procedures. Temporary uses and structures are currently getting much attention, and as the Turin case demonstrates, they are still considered as an alternative, unconventional and informal approach to development. The Turin case is an example of urban reuse: an old building was transformed into a modern and functional one, with new functions and meanings. Temporary projects seem relevant in exploring alternative scenarios and reacting to instant challenges. They often could present alternatives for filling the voids left over by failed policies.

Social housing represents a new housing paradigm that calls into question the policies of architectural design and urban development that have characterised past decades. The new cultural model brought to life through these housing projects seeks to move beyond logic focused in particular on meeting the housing needs of certain social groups. Instead, the sustainable projects of social housing connect urban regeneration with economic development and involve a range of different actors and sets of knowledge and skills in synergistic collaboration. The search for urban quality also undoubtedly constitutes a key issue in disciplinary discussions of urban regeneration issues. This space offers significant opportunities for planning, urban renewal and regeneration [3].

The analysis of the selected case studies sought to demonstrate how spatial structure and social structures are strictly interconnected to create social innovation and social sustainability.

Housing is therefore considered as a multidimensional process that involves different actors (institutions, cooperatives, agencies (private and public); different social groups, planners and local stakeholders) who interact by sharing objectives, methods and strategies.

The range of actors involved and the integration of economic, social and cultural aspects, it is the individual in all his or her complexity who experiences specific needs through a housing system that also includes other aspects of life. The cases analysed seek to create new housing models and solutions that contribute to improving the socio-economic fabric of the local area. In the Turin case, for example, there is an effort to plan, organise and collectively manage the socially useful services related to social housing facilities such as, for example, after-school programmes for children. However, this initial appraisal of the projects carried out in Italy does reveal that, unlike projects in other European countries, Italian public administrations continue to tackle the housing problem using a sectorial approach. Social housing instead increasingly calls for a multidimensional approach in which the needs and requirements of individuals, communities and society are concretely intertwined with new forms of market and governance involving financial, real estate, technological and social expertise. As these two case studies show, the range of actors involved and the integration of economic, social and cultural aspects, it is the individual in all his or her complexity who experiences specific needs through a housing system that also includes other aspects of life. These cases seek to create new housing models and solutions that contribute to improving the socio-economic fabric of the local area. In the Turin case, for example, there is an effort to plan, organise and collectively manage the socially useful services related to social housing facilities such as, for example, after-school programmes for children. In the experience of Zoia social housing in Milan, the cooperatives assigned through a competition Officine Creative spaces to the following subjects: a flautist, a group of women scenographers, an art gallerist and two architects also engaged in the design of a series of activities addressed to the inhabitants and the neighbourhood. In

Milan, children's music education courses or sports initiatives create 'an architecture of relationships' that represents one of the main aims of social housing [25, 26].

As regard to the governance models, the analysis of several cases in Italy and Europe shows that more coordination is needed with public institutions, as they can contribute positively to the success of projects thanks to the joint formulation of housing policies and services. New options financed by public/private institutions that are dedicated to social housing are emerging. They represent innovative solutions for structuring, reusing and managing social and affordable housing projects that are economically sustainable and not dependent on grants. 'These new models have not only opened the social rental sector to private and public/private investments but have also demonstrated how a virtuous intersection of the three crucial policy dimensions—housing, urban and social policies—can help redraw the boundaries of local welfare' [27]. In Europe, studies and researches highlight the emerging demand of houses and at the same time the demand of housing services: offering a house means not only producing efficient buildings but providing people with tools to improve the quality of life.

Social housing is not yet diffused in Italy. However, the experiments that are already operational do allow us to identify three main directions for new housing as 'living policies'.

First of all, people, with their specific needs and aspirations, must be placed once again at the centre of analytical thinking and contemporary design [28–34].

Secondly, living spaces need to enter into dialogue with their surrounding habitats, a relationship often neglected in housing policy that is more oriented towards quantity than quality and the market. Thirdly, it is increasingly crucial that social housing cover not only the 'grey segment' of the population but also the most marginal strata, now more than ever facing conditions of serious hardship. Tackling the housing emergency also requires greater investment in measures that not only are temporary and flexible but also ensure stable accommodations as well as job-training and employment placement assistance to the most underprivileged individuals; above all, they should provide permanent solutions. Regarding the possibility to realise in practice housing policies that can include social innovation and environmental sustainability, it seems to be an exciting direction, actually restricted in Italy to successful but very limited experiences.

Author details

Rossana Galdini and Silvia Lucciarini*
Department of Social and Economic Sciences, Sapienza University of Rome, Italy

*Address all correspondence to: silvia.lucciarini@uniroma1.it

References

[1] Gregotti V. Architettura e postmetropoli. Torino: Einaudi; 2011

[2] De Carlo G. Questioni di architettura e di urbanistica. Santarcangelo di Romagna (RN): Maggioli; 2010

[3] Galdini R. L'abitare difficile. In: La casa in Italia tra desideri e risorse. Liguori, Napoli; 2012

[4] Intervista a Camilo Magni – Architetti senza Frontiere Italia. https://www.industriarchitettura.it/2016/09/21/intervista-a-camillo-magni-architetti-senza-frontiere-italia/

[5] Clapham D. The Meaning of Housing: A Pathways Approach. Bristol: The Policy Press; 2006

[6] Hickman P, Robinson D. Transforming Social Housing: Taking Stock of New Complexities. HousingStudies. 2006;**21**(2):157-170

[7] The State of Housing in the EU 2017 Housing is still Europe's challenge Brussels, 17 October 2017 | Published in Energy, Urban, Economy, Social, Research www.housingeurope.eu

[8] Galdini R. Emergenza abitativa e pratiche informali. Il caso di Roma. In: Sociologia Urbana e Rurale, FrancoAngeli, Milano, Anno XXXVIII, n.112/2017. 2017

[9] Tosi A. Le case dei poveri. È ancora possibile pensare un welfare abitativo? Milano: Mimesis; 2018

[10] Santangelo S. Politiche per la casa: Una difficile transizione. Roma: Carocci; 2015

[11] AdminHow to Become a Social Architect: An Interview with Alejandro Haiek; 2016 . http://masteremergencyarchitecture.com/author/ad

[12] EUROSTAT, Living Condition in Europe 2018 Edition. https://ec.europa.eu/eurostat/documents/

[13] Czischke D. Social Innovation in Housing: Learning from practice across Europe' Discussion paper commissioned by the Chartered Institute of Housing on behalf of the Butler Bursary. The Netherlands: Delft University of Technology; 2013. http://www.cih.org/resources/PDF/Membership/Social%20 Innovation%

[14] Crudeli A. Alejandro Aravena e l'archiettura del ribaltamento, dic. In: Architettura, urbanistica. 2016. https://www.centoventigrammi.it/alejandroaravena/

[15] Boano C, Vergara Perucich F. Architettura felice a metà; 2016. http://www.zeroundicipiu.it/wp-content/uploads/2016/04/VV04_ITA_091.p

[16] Ferreira JSW, Ferrara L. "The Formulation of a New Urban Matrix in Brazil, Basedon Socio-Environmental Justice." 3 vols. Vol. 3: Habitação social e sustentabilidade of Sustentabilidade urbana: Impactos do desenvolvimentoeconômico e suasconseqüênciassobre o processo de urbanizaçãoempaísesemergentes: Textos para asdiscussões da Rio+20. Brasília: Ministério do Meio Ambiente; 2015. pp. 55-92

[17] Cavalcante M. Lis. Re-thinking the urban form of social housing in Brazil: A search for urban sustainability. [thesis]. Rochester Institute of Technology; 2016. Accessed from: https://scholarworks.rit.edu/cgi/viewcontent.cgi?referer=https://www.google.com/&httpsredir=1&article=10363&context=theses

[18] Manzi T, Lucas K, Lloyd J, Allen J. Social Sustainability in Urban Areas, Communities, Connectivity and the

Urban Fabric. London, Washington, DC: Earthscan; 2010

[19] Colantonio A. Social sustainability: A review and critique of traditional versus emerging themes and assessment methods. In: Horner M, Price A, Bebbington J, Emmanuel R, editors. Sue-Mot Conference 2009: Second International Conference on Whole Life Urban Sustainability and Its Assessment: Conference Proce. Loughborough: Loughborough University; 2009. pp. 865-885

[20] Egan J. The Egan Review: Skills for sustainable Communities, Paperback, April 15. London: Riba Enterprises; 2004

[21] Florida R. The New Urban Crisis: How Our Cities are Increasing Inequality, Deepening Segregation, and Failing the Middle Class — and What We Can Do About It. New York: Basic Book; 2017

[22] Harvey D. Rebel Cities: From the Right to the City to the Urban Revolution. London: Verso Books; 2012

[23] Balducci A. Pianificazione strategica e politiche di sviluppo locale. Una relazione necessaria? In: Archivio di Studi Urbani e Regionali. Vol. 64. 1999

[24] Donzelot J. Faire société. Paris: Seuil; 2003

[25] Benetti A. Progetto Zoia. Social Housing a Milano; 2016. https://www.artribune.com/progettazione/architettura/2016/04/progetto-zoia-social-housing-periferia-milano-federica-verona-architettura/

[26] Lucciarini S. Ruolo e rapporti tra attore pubblico e privato nel governodel territorio. Il caso di Roma in Sociologia Urbana e Rurale 116; 2018

[27] Rizzica C. New social and affordable housing. 2017. https://www.friendsofeurope.org/publication/new-social-and-affordable-housing

[28] Czischke D. Collaborative housing and housing providers: Towards an analytical framework of multi-stakeholder collaboration in housing co-production. International Journal of Housing Policy. 2017:1-27

[29] Ferri G. Cambiare l'abitare cooperansdo. Milano: Mondadori; 2018

[30] Moulaert F, Martinelli F, Swyngedouw E, Gonzalez S. Can neighborhoods save the city? Community development and social innovation. London, New York: Routledge; 2010

[31] Simms JR. Technical and social innovation determinants of behavior. Systems Research & Behavioural Science. 2006;**23**(3):383-393

[32] European Union. The Young Foundation (2010) Study on Social Innovation. A paper prepared by the Social Innovation eXchange (SIX) and the Young Foundation for the Bureau of European Policy Advisors

[33] Sustainable House. http://www.gf.uns.ac.rs/~wus/wus09/Sustainable%20House%20web%20page.doc%202.htm

[34] Triveno L. The "starchitect" of the poor: The keys to Alejandro Aravena's work, Sustainable Cities. 2016. http://blogs.worldbank.org/sustainablecities/starchitect-poor-keys-alejandro-aravenas-work?CID=ECR_LI_worldbank_EN_EXT

[35] Taylor Ramsey J. The Validity of Historical Narrative, Seattle pacific University Spring June 7th. 2017. https://digitalcommons.spu.edu/honorsprojects

[36] King P. Housing Boom and Bust. Owner occupation, governement

regulation and the credit crunch. London, New York: Routledge; 2008

[37] Spinozzi P, Mazzanti M. Cultures of sustainability and wellbeing theories. Histories and Policies. London, New York: Routledge; 2018

Chapter 6

Socioeconomic Influences on Affordable Housing Residents: Problem Definition and Possible Solutions

Deborah J. Bowen and Lisa Quintiliani

Abstract

Socioeconomic status (SES) is a powerful social determinant of health. Often, affordable housing is an important step in promoting reliable economic and social health among individuals living in poverty. However, we argue that we must go further to improve the long-term health outcomes of these individuals and families. First, we use survey data and geographical analysis to identify the socioeconomic status of neighborhoods and residents of affordable housing in a major urban center. SES levels are certainly lower among affordable housing residents, and SES was significantly lower in public housing development neighborhoods than other neighborhoods. We offer solutions from our own and other research experiences that identify potential changes to affordable housing to promote and maximize health of residents. These data have implications for multilevel intervention.

Keywords: built environment, neighborhood, socioeconomic status, affordable housing

1. Introduction

Public housing residents have a much higher burden of chronic disease and disease risk than other urban dwellers, even when controlling for income [1]. One possible driver of these differences may be the environment in which public housing developments are located [2]. There is growing evidence that the physical surroundings and resources available to individuals can have an impact on the relationship between socioeconomic status (SES) and health outcomes [3–5].

Early research led to several large-scale interventions that provide alternatives to living in densely populated, very low SES neighborhoods and buildings [6]. The Moving to Opportunity study used a randomized design to demonstrate that health outcomes, including obesity, could be improved when moving from high- to low-poverty areas. Many of these interventions and changes required physical changes to public and affordable housing that take years to create and are expensive and labor intensive. In contrast, we believe that with simple changes, we can create affordable housing experiences that promote health, rather than reduce the likelihood of health promotion efforts. This chapter articulates those potential experiences and calls for research into them.

IntechOpen

This chapter is divided into two main sections. First, we report health indicators and the socioeconomic status of residents and of neighborhoods within public housing developments in Boston, Massachusetts, and compare it with the SES levels in other Boston neighborhoods. This landscape is likely to be found in all major urban settings in the United States. Then, we propose methods of ameliorating the effects of SES on the health of affordable housing residents and discuss the literature to support these methods.

2. SES and affordable housing

2.1 Setting

This research was conducted as part of the US Centers for Disease Control and Prevention-funded Prevention Research Center at Boston University. Research at this center is focused on 69 public housing developments in Boston with almost 30,000 residents (>18,000 adults) in 13,937 units or apartments, with 68% of adult residents as female. On average, PHDs have greater than 300 units. The most common ethnic group in the PHDs is Hispanic (35%), followed by Black (32%) and White (21%). English (52%) and Spanish (33%) are the primary languages spoken in PHDs. Average annual household income is $13,700, which is below the poverty line for a household with only one adult and one child.

2.2 Study 1: SES differences between housing residents and other urban dwellers

First, we analyzed data from an ongoing government-sponsored survey in order to compare the health of public housing residents with other Boston residents through this random-digit-dial survey. We used data from the Boston Behavioral Risk Factor Surveillance System collected in 2001 and 2003 to make crude and demographically adjusted comparisons between public housing residents and other Boston city residents on measures of health status, access and utilization, and health behaviors [1].

Public housing residents were more likely to report fair or poor overall health status, ever-diagnosed hypertension, current asthma, ever-diagnosed diabetes, obesity, disability, loss of six or more teeth, and feelings of depression for 15 days or more in the past month. Public housing residents were slightly more likely than others to be without health insurance or report financial barriers to medical care. Public housing residents reported more smoking and physical inactivity, less past-month binge drinking and past-year marijuana use, and similar levels of lifetime drug use.

This remarkably consistent evidence convinced our research group to focus our efforts on public housing residents in Boston and elsewhere. We focused specifically on obesity and smoking as two key behaviors that are amendable to change but that are double to triple the rates among public housing residents compared with other urban dwellers, given their role in promotion of chronic disease rates in the United States [7, 8]. Our smoking research has taken the form of development and resident-based smoking policies that restrict or limit smoking exposure among all residents of public housing [9]. We now promote smoke-free housing across the country and assist other public and affordable housing settings to go smoke-free [10]. The Boston Housing Authority initiated their smoke-free housing policy (not smoking in resident apartments) in 2012.

2.3 Study 2: analysis of neighborhoods surrounding public housing developments

Our focus on obesity was on understanding the risk factors for obesity in public and affordable housing and in testing new interventions to reduce obesity in those settings. The second source of data was from the publically available 2007–2011 American Community Survey (ACS) summary file. The American Community Survey is an ongoing survey administered by the United States Census Bureau that collects a variety of demographic and economic data [11]. Block groups containing a family public housing development were considered public housing neighborhoods.

Several measures of SES were taken from the ACS, using the 2010 block group as the unit of analysis: median household income (in dollars); median value of owner-occupied housing (in dollars); percentage of households receiving interest, dividend, or net rental income; percentage of adults 25 years or older who completed high school; percentage of adults 25 years or older who completed college; percentage of employed persons 16 years or older in executive, managerial, or professional occupations; percentage of persons below the US poverty line, percentage white race, and percentage unemployed. To calculate area-level SES for each block group represented, we used the method of Diez-Roux [12], a combination of six census-based variables at the block group level. The Gini coefficient measures the inequality among values of a frequency distribution and is used here as a measure of income inequality [13].

Table 1 contains the socioeconomic levels of neighborhoods of Boston's family public housing developments compared with neighborhoods in which there was not a family development. As seen in this table, there were consistent and significant differences in socioeconomic status variables between development neighborhoods and nondevelopment neighborhoods. This included the average Diez-Roux score for development and nondevelopment neighborhoods and for all of the specific variables included in the composite score, except for value of owner-occupied housing. The

	Public housing (N = 26)	**Nonpublic housing (N = 609)**	**P-value**
Characteristic	**Mean (SD)**	**Mean (SD)**	
Mean Diez-Roux composite score	−4.01 (4.48)	0.18 (4.50)	<0.0001
SES indicators in composite score			
Median household income ($)	28,513 (22,066)	59,088 (32,179)	<0.0001
Median value of owner-occupied housing ($)	372,308 (132,679)	395,254 (152,733)	0.0592
% households receiving interest, dividend, or net rental income	9.75 (12.19)	22.15 (14.31)	<0.0001
% adults 25 years or older who completed high school	71.62 (15.95)	84.09 (14.25)	<0.0001
% adults 25 years or older who completed college	25.01 (25.24)	39.45 (26.84)	0.007
% employed persons 16 years or older in executive, managerial, or professional occupations	30.07 (18.70)	41.76 (21.68)	0.007
Additional area-level SES indicators			
% persons below 100% of federal poverty level	42.49 (19.41)	19.97 (17.07)	<0.0001
% white race	37.16 (20.91)	56.80 (31.97)	0.002
% unemployed (imputed from Census 2010 tract level)	13.92 (6.59)	10.30 (7.93)	0.034
Gini index of income inequality (imputed from Census 2010 tract level)	0.498 (0.069)	0.461 (0.082)	0.033

Table 1.
Mean block group characteristics by public housing status (N = 635 2010 census block groups).

differences between development and nondevelopment neighborhoods were also found in percentage of persons living below the poverty line, percentage unemployed, and percentage white race. The average Gini index measure of income inequality was higher in development neighborhoods than nondevelopment neighborhoods.

Figure 1 is a map of Boston with family developments outlined in red and the block groups within Boston color-coded by the Diez-Roux composite measure of socioeconomic status. Many of the public housing development neighborhoods are in block groups color-coded in white or lighter colors, indicating a low composite score of socioeconomic status. Boston is a city with extensive public transportation

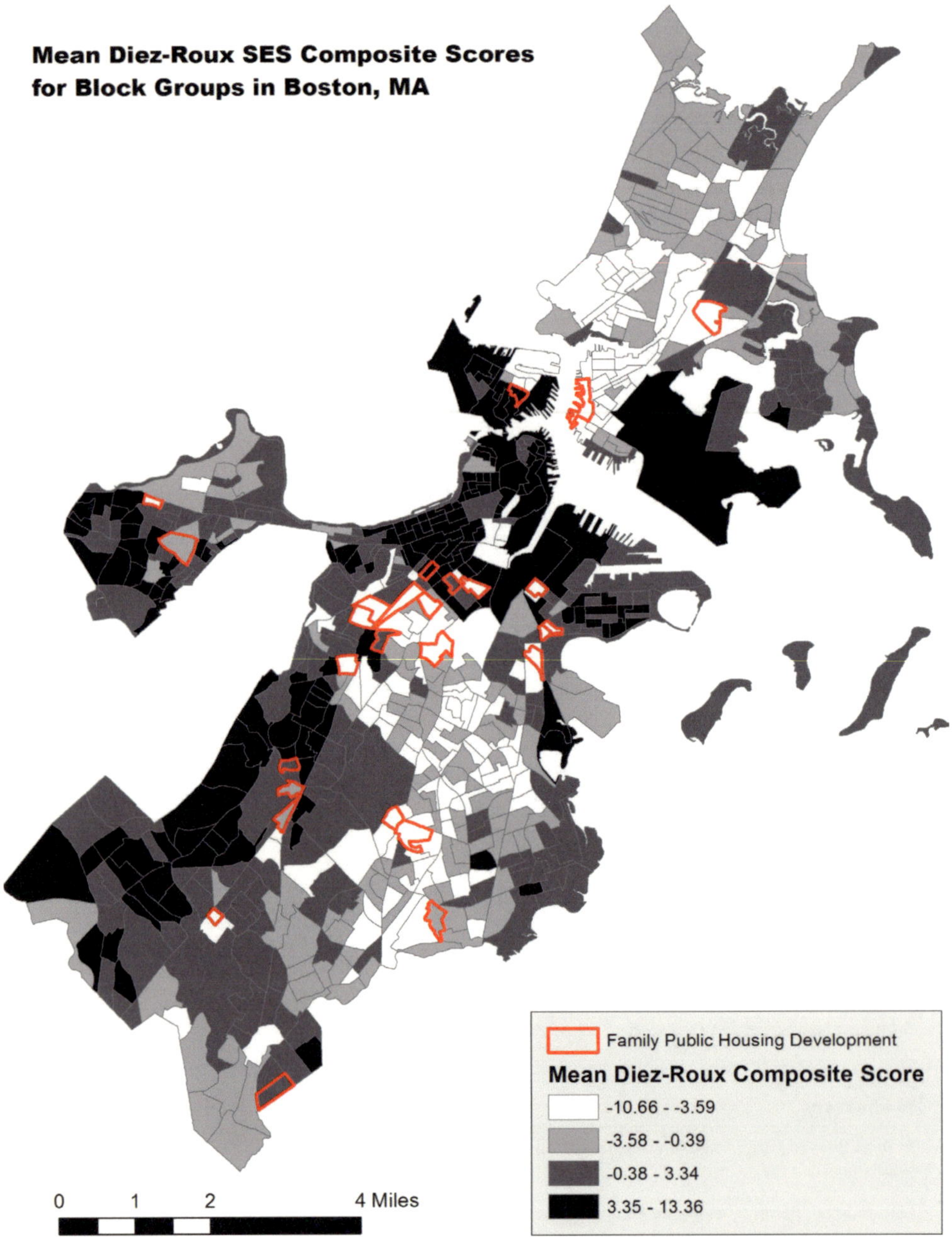

Figure 1.
Map of Boston, MA, showing the socioeconomic status of public housing neighborhoods and nonpublic housing neighborhoods using an index calculated from 2007 to 2011 American Community Survey data.

where many children are bused to school from across town and workers can also readily access the transit system, so there is opportunity for movement across the city by public housing residents. Nevertheless, many people use services and have social interactions within a quarter mile of their home or work. Given that most residents of public housing are underemployed, resources built into the lived neighborhood become even more important.

We can hypothesize that these SES differences are likely associated with different amounts of opportunity for healthy eating and activity choices, both related to obesity. The fundamental distinctions in SES associated with differential access to healthy opportunities still exist [14, 15]. The concept of differential access within the practical confines of one's neighborhood frames what the next steps are for intervention to change an unhealthy, obesogenic environment.

2.4 Study 3: changing the environment of public housing

Given our thinking around the utility of simple changes within or in the immediate surrounding area of public housing developments, the third study was an initial test of an intervention to improve the environments of public housing developments to reduce obesity. The Healthy Families study [16, 17] was a group randomized weight reduction trial, in which five housing developments were randomized to receive all physical and social environment intervention activities and five to an assessment only control group. Intervention activities were promoted by resident advocates and targeted weight-related behaviors according to multiple levels: physical environment (e.g., walking groups, resource maps), social environment (e.g., health screenings, cooking demonstrations), consumer (mobile food bus), and information environment (social media). A cohort of 211 (intervention = 116; control = 95) randomly selected women and daughter pairs formed the evaluation cohort. Overall, most participants were Latino (63%), had affordable health insurance (79.1%), and have a high school education or less (64%). The 1-year follow-up assessment data indicated significant effects on moderate-level physical activity; these data were used, in part, to calculate power for the present study. Briefly, physical activity increased from 19.8 minutes per day to 30.5 minutes per day for intervention women, while control women reported stable walking levels. The proportion of sedentary intervention women decreased from 89 to 59%; again control women remained stable. These data indicate that a development-wide intervention can make a difference in the activity levels of affordable housing residents. The changes in outcomes were profound, in that a very large proportion of women changed from sedentary to nonsedentary in 1 year. This study showed that with relatively easy improvements in the environmental cues for healthy behaviors, public housing residents could make changes that in the short term can result in decreased obesity and decreased sedentary behaviors.

3. Future research and opportunities for affordable housing residents

One lesson from the previously cited Healthy Families intervention study is the idea that residents themselves, with help and guidance, can help other residents to improve their healthy behaviors and reduce obesity and smoking behaviors. We have conducted multiple projects to train and support residents to become lay health workers within public and affordable housing, skilled in behavior change techniques, motivational discussions with other residents, and environmental changes [18, 19]. In Healthy Families, for example, residents were trained and supported to work within their own and others' developments to help other residents participate in walking groups [20] and choose more healthful food offerings. These lay health workers, called Healthy Living Advocates, participated in the design and implementation of all intervention activities within Healthy Families, and many are still active in their developments several years later.

Training HLAs in nutrition and physical activity changes skills on a large scale requires a curriculum that can be easily delivered and transported from setting to setting and delivered using easy to use accessible methods for affordable housing residents. eHealth technologies offer such a platform. Our survey data support the idea that recent use of the Internet and social media is high among public housing residents, with 65% and 59% reporting past-day use, respectively. Almost all residents in the sample used their cell phone for at least one phone call daily (97%) and 84% reported receiving or sending at least one text message per day [21].

Therefore, the use of eHealth technology may be used in several ways to facilitate the delivery of a curriculum by lay health workers: to provide training, help guide conversations with fellow residents, and provide a place to store notes about each contact for later retrieval by the lay health worker and the research team overseeing quality assurance. We have created an eHealth website that consists of web pages for use by lay health workers to counsel individuals in-person or over the telephone about diet and physical activity behaviors. The website assists the lay health worker by displaying on their computer screen [1] previously collected data about the resident receiving counseling (obtained via a survey assessment) and [2] suggested questions lay health advisor can ask the person about his/her lifestyle to identify relevant social contextual factors (e.g., family roles) that impact the diet and physical activity behaviors, followed by suggestions on how the lay health worker can work with the person to set behavior change goals (if the participant is motivated/ready). The website then directs behavior-specific goals which are then set with the participant and recorded in the website, and suggestions for how to meet the goals are discussed (e.g., enroll the help of a support person). All text is written in a motivational interviewing style and provides cues to the CHW to provide reflections (both simple and complex). The lay health advisor records notes within the website for their use during the next session and for the researcher's use in understanding what happened during the counseling session.

The website has undergone two rounds of usability testing with lay health advisors from public housing developments. Five lay health advisors reviewed an early version of the website and provided qualitative comments on its design and rated its usefulness and ease of use highly (mean = 6.8, on a 7-point scale). Second, we conducted another round of testing with three resident lay health advisors (two new individuals, one who had participated in round 2). Over 6 months, we conducted an iterative development plan in which we had eight one-on-one meetings to show them our plans, obtain their feedback, implement these changes, and then show them newly revised versions. They rated the CuesWeight system highly in terms of usefulness and ease of use (average of 1.7 on a scale of 1 [best] to 7 [worst]). Overall, these findings support the usefulness of an eHealth website to deliver a curriculum that can be easily delivered and transported from setting to setting and delivered using easy to use accessible methods for affordable housing residents.

4. Conclusions

Public and affordable housing residents need the help of research and public health practice teams. We have taken the approach of collaborating with the Partnership in Health and Housing to work with community members to develop and implement survey and intervention work. For academic institutions without an existing partnership, researchers and public health practice teams can forge relationships with the tenant task forces that are present in many public and affordable housing communities.

With relatively simple changes, we can make potential differences in the health outcomes of public and affordable housing residents [22]. For example, using eHealth technologies as a platform for training and delivering lay health advisors in public housing developments is an opportunity to expand the reach of workers both within their developments and across developments nationally. Websites could be made open source so that lay health advisors could review and modify the content to fit their particular populations, housing contexts, and targeted health behaviors.

Changing policy to fund these types of programs and activities will improve the systems that support residents and help them improve their health. Funding positions of community health workers could be supported along with regular maintenance of the physical building to enable residents to improve their health outcomes and live happier, healthier lives.

Acknowledgements

This article is a product of a Prevention Research Center and was supported by Cooperative Agreement Number 5 U48 DP0019-22 from the Centers for Disease Control and Prevention. The findings and conclusions in this journal article are those of the author(s) and do not necessarily represent the official position of the Centers for Disease Control and Prevention. This work was also supported by a Scientist Development Grant (14SDG20050015) from the American Heart Association.

Author details

Deborah J. Bowen[1*] and Lisa Quintiliani[2]

1 Bioethics and Humanities, University of Washington, Seattle, WA, USA

2 Boston University School of Medicine, USA

*Address all correspondence to: dbowen@uw.edu

References

[1] Digenis-Bury EC et al. Use of a population-based survey to describe the health of Boston public housing residents. American Journal of Public Health. 2008;**98**(1):85-91

[2] Anderson LM et al. Providing affordable family housing and reducing residential segregation by income. A systematic review. American Journal of Preventive Medicine. 2003;**24**(3 Suppl):47-67

[3] RE L, Mama SK, McAlexander KP, Adamus H, Medina AV. Neighborhood and PA: Neighborhood factors and physical activity in African American public housing residents. Journal of Physical Activity & Health. 2011;**8**(Suppl 1):S83-S90

[4] Caspi CE et al. The relationship between diet and perceived and objective access to supermarkets among low-income housing residents. Social Science & Medicine (1982). 2012;**75**(7):1254-1262

[5] Caspi CE et al. The social environment and walking behavior among low-income housing residents. Social Science & Medicine. 2013;**80**:76-84

[6] U.S. Department of Housing and Urban Development. Expanding Housing Choices for HUD-Assisted Families: First Biennial Report to Congress. 1996. Available from: http://www.huduser.org/portal/publications/affhsg/choices.html [Accessed: August 14, 2010]

[7] Field AE, Coakley EH, Must A, Spadano JL, Laird N, Dietz WH, et al. Impact of overweight on the risk of developing common chronic diseases during a 10-year period. Archives of Internal Medicine. 2001;**161**(13):1581-1586. DOI: 10.1001/archinte.161.13.1581

[8] Whitlock G, Lewington S, Sherliker P, Clarke R, Emberson J, Halsey J, et al. Body-mass index and cause-specific mortality in 900 000 adults: Collaborative analyses of 57 prospective studies. Lancet. 2009;**373**(9669):1083-1096. DOI: 10.1016/S0140-6736(09)60318-4

[9] Brooks DR, Burtner JL, Borrelli B, Heeren TC, Evans T, Davine JA, et al. Twelve-month outcomes of a group-randomized community health advocate-led smoking cessation intervention in public housing. Nicotine & Tobacco Research. 2018;**20**(12):1434-1441

[10] Available from: www.BuildingSuccessSmokefree.org

[11] United States Census Bureau, American FactFinder. American FactFinder. Available from: http://factfinder2.census.gov/faces/nav/jsf/pages/searchresults.xhtml?refresh=t; http://www.census.gov/acs/www/ [Accessed: August 10, 2014]

[12] Diez-Roux AV et al. Area characteristics and individual-level socioeconomic position indicators in three population-based epidemiologic studies. Annals of Epidemiology. 2001;**11**(6):395-405

[13] Yen IH, Kaplan GA. Poverty area residence and changes in physical activity level: Evidence from the Alameda County Study. American Journal of Public Health. 1998;**88**(11):1709-1712

[14] Larson NI, Story MT, Nelson MC. Neighborhood environments: Disparities in access to healthy foods in the U.S. American Journal of Preventive Medicine. 2009;**36**(1):74-81

[15] Hilmers A, Hilmers DC, Dave J. Neighborhood disparities in access to healthy foods and their effects on environmental justice.

American Journal of Public Health. 2012;**102**(9):1644-1654

[16] Quintiliani LM, DeBiasse MA, Branco JM, Bhosrekar SG, Rorie JA, Bowen DJ. Enhancing physical and social environments to reduce obesity among public housing residents: Rationale, trial design, and baseline data for the healthy families study. Contemporary Clinical Trials. 2014;**39**(2):201-210

[17] Bowen DJ, Quintiliani L, Goodman R, Smith E. Changing the housing environment to reduce obesity in public housing residents: A cluster randomized trial. Vol. 18. BMC Public Health; 2018. Article number: 883

[18] Bowen D, Battaglia T, Murrell S, Bhosrekar S, Caron S, Smith E, et al. What do public housing residents say about their health? Progress in Community Health Partnerships: Research, Education, and Action. 2013;**7**(1):39-47

[19] Bowen DJ, Bhosrekar SG, Rorie J, Goodman R, Thomas G, Maxwell N, et al. Resident health advocates in public housing family developments. Family & Community Health. 2015;**38**(2):141-148

[20] Bowen DJ, Rus A, Beltrane C, Drayton M, Williams MJ, Goodman R. Walking groups: A simple, affordable intervention program for public housing developments. In: Brennan V, Kumanyika S, Zambrana, editors. Obesity Interventions in Underserved Communities. Baltimore: Johns Hopkins University Press; 2014

[21] Quintiliani LM, Reddy S, Goodman R, Bowen DJ. Information and communication technology use by female residents of public housing. Mhealth. 2016;**2**:39

[22] Bowen DJ, Barrington WE, Beresford SAA. Identifying the effects of environmental and policy change interventions on healthy eating. Annual Review of Public Health. 2015;**36**:289-306